House wiring
SIMPLIFIED

tells and shows you how

FLOYD M. MIX
President
The Goodheart-Willcox Co., Inc.

South Holland, Illinois
THE GOODHEART-WILLCOX CO., Inc.
Publishers

INTRODUCTION

HOUSE WIRING SIMPLIFIED teaches the fundamentals of modern house wiring procedures.

It shows, in easy-to-understand drawings, how to install wiring that will serve SAFELY and EFFICIENTLY, the Lighting, Appliance and Equipment needs of Today, as well as Tomorrow.

HOUSE WIRING SIMPLIFIED tells how to determine what wiring is needed. It covers both new homes and older homes.

HOUSE WIRING SIMPLIFIED is intended for students in high school, vocational school, college, apprentice training and adult classes. It will enable the home owner to handle many wiring jobs he was reluctant to tackle before.

Floyd M. Mix

House Wiring Simplified

CONTENTS

House Wiring Simplified

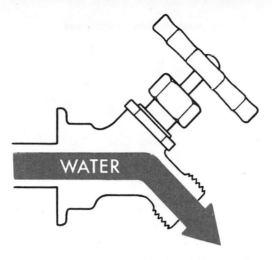

Fig. 1-1. Water flows through a pipe; electricity flows through a wire.

Unit 1

ELECTRICAL TERMS

In House Wiring, we make frequent reference to the terms VOLTS, AMPERES, WATTS.

It is easy to understand these terms, if we compare the flow of electricity through a wire, to the flow of water through a pipe. See Fig. 1-1.

VOLTS (V)

Water pressure is measured by pounds; electric pressure by VOLTS.

AMPERES (amps.)

In measuring the amount of water passing through a pipe, the term gallons is used. In electricity, the unit of measurement is AMPERES. Various components (parts) of a wiring system .... fuses, wall switches, outlets, etc., are rated in amperes. The rating of the components in amperes, indicates the greatest amount of current with which they should be used.

WATTS

The WATT is the unit of measurement that tells us how much electricity is being used. Watts consumed are determined by using this formula:

$$\text{Watts} = \text{Volts} \times \text{Amperes}$$

KILOWATT: 1,000 watts.

KILOWATT HOUR (KWH): 1,000 watts used for one hour. The KILOWATT HOUR is the unit by which electricity is metered (sold).

CIRCUIT

In an ELECTRICAL CIRCUIT, a wire is provided for electricity to flow from the point of supply (generator or distribution box) to the final use such as an electric light, or appliance, and back, Fig. 1-2.

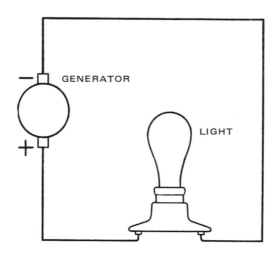

Fig. 1-2. A simple electrical circuit.

RESISTANCE

RESISTANCE may be described as electrical friction or the tendency of a conductor (wire) to keep the electric current from passing through it. Electrical energy lost is given off as heat. Conductors such as copper, silver, and aluminum, offer very little resistance to the flow of electric current. Examples of poor conductors (insulators) are glass, wood, paper.

OHMS

The OHM is the unit of measurement of electrical resistance. We speak of ohms of resistance in electricity like we speak of pounds of pull required to break a certain fishing line.

SWITCH

A SWITCH is a device for controlling the flow of current in an electric circuit by opening and closing the circuit, Fig. 1-3. With the switch in the open position there is no electrical connection between the terminals.

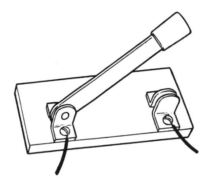

Fig. 1-3. Knife type switch, for low voltage only, in open position (open circuit, no electrical flow).

DIRECT CURRENT (dc)

DIRECT CURRENT flows only in one direction. Batteries (such as storage and dry cell) are important sources of direct current. One terminal of a battery is always positive (+) and the other is always negative. (-).

ALTERNATING CURRENT (ac)

In ALTERNATING CURRENT the voltage flows first in one direction, then the other. Each two reversals of direction of current flow is called a cycle. The number of cycles per second is called FREQUENCY. There is a very short interval of time between changes of direction when no current is flowing.

Most house wiring in the United States is 60 cycle. You may wonder why an electric light operated by current which changes direction of flow 60 times per second does not flicker. This is because the filament in the light bulb does not have time to cool while there is no current flowing.

PROGRESS CHECK — UNIT 1

NOTE: Progress Checks have been included in this book for these reasons:

1. They will help you determine what you have learned, also what material you should restudy.

2. They summarize for you many important items with which you should become familiar.

You can see if your answers are correct, by referring to the Answer Keys, starting on page 171. Be fair with yourself. Answer the questions (on a separate sheet of paper) before you refer to the correct answers in the Answer Key.

Matching quiz. Match the terms in the column at the left with the proper definition on the right:

a. Alternating current.
b. Amperes.
c. Direct current.
d. Kilowatt hour.
e. Ohms.
f. Resistance.
g. Switch.
h. Volts.
i. Watts.

1._____ Measurement of electrical pressure.
2._____ Measurement of amount of electricity.
3._____ Measurement of how much electricity is being used.
4._____ Unit by which electricity is metered.
5._____ Measurement of electrical resistance.
6._____ Tendency of conductor to keep electric current from passing through it.
7._____ Device for controlling current flow by opening or closing circuit.
8._____ Current flow in one direction only.
9._____ Current flowing first in one direction, then in other.

Unit 2

CONDUCTORS

In house wiring, electrical conductors which provide paths for the flow of electric current, are wires over which an insulating material is formed. Insulation is a noncurrent-carrying material which insures that the current flow will be through the wire.

WIRE SIZE

The drawing, Fig. 2-1, shows actual size of conductors, without insulation.

Note that, as the numbers become larger, the size of the wire decreases. For most house wiring jobs, copper wire numbers 12 and 14 are specified by the building plans. Numbers 6 and 8 wires, which are available either as solid wire, or stranded, are used for heavy power circuits, and as service entrance leads into buildings.

Current-carrying capacities of conductors of various sizes are shown in Fig. 2-2.

In house wiring, considerable use is made of single wire conductors, Fig. 2-3.

There are many installations where the use of individual wires spaced and supported side by side would be inefficient and impractical. For these installations,

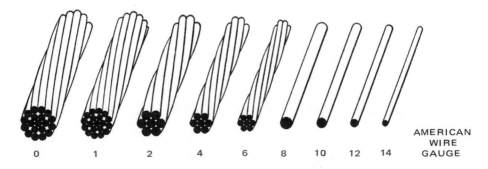

Fig. 2-1. Copper conductors by gauge number. The wire gets smaller as the gauge numbers get larger.

conductor cables consisting of insulated wires arranged in pairs or groups of three or more, are used. Additional insulating or protective material is formed or wound around the insulated wires.

	SIZE		CURRENT-CARRYING CAPACITY (Amperes)			
Number	Diameter (Inches)	Weight (Feet Per Pound)	Rubber-Insulated Wire In Conduit Or Cable	Rubber-Insulated Wire On Insulators	Weather-Proof Wire On Insulators	Resistance (Ohms Per 1,000 Feet)
24	.0201	817.6	— —	— —	— —	— —
18	.0403	203.4	— —	— —	— —	— —
16	.0508	127.9	— —	— —	— —	— —
14	.0640	80.44	15	24	30	2.48
12	.0808	50.59	20	31	39	1.56
10	.1018	31.82	25	42	54	0.98
8	.1284	20.01	35	58	71	0.62
6	.184	12.58	50	78	98	0.39
4	.232	7.91	70	105	130	0.24
2	.292	4.97	90	142	176	0.15
1	.332	3.94	100	164	203	0.12
1/0	.373	3.13	125	193	237	0.10
2/0	.419	2.48	150	223	274	0.08
3/0	.470	1.97	175	259	318	0.06
4/0	.528	1.56	225	298	368	0.05

Fig. 2-2. Copper conductors: diameter, weight, current-carrying capacity in amperes, resistance in ohms per 1,000 feet.

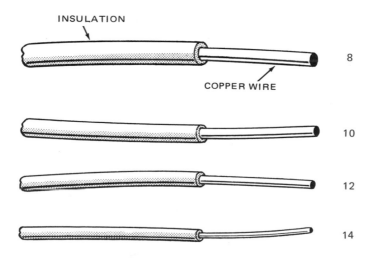

Fig. 2-3. Single wire conductors.

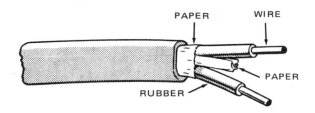

Fig. 2-4. Two-wire nonmetallic sheathed cable without ground.

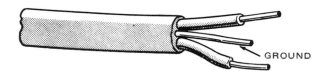

Fig. 2-5. Two-wire sheathed cable with ground wire.

In a two-wire cable, one wire is black, one white. In a three-wire cable the extra wire is red. One wire (always the white wire) will be a neutral or ground wire, the colored wires with the exception of a grounding wire (green) as discussed later will be "hot" wires. Under certain conditions grounding wires may be bare and uninsulated.

NONMETALLIC, ARMORED CABLES

In modern house wiring, two types of cable are commonly used, NONMETALLIC SHEATHED CABLE which is an assembly of two or more insulated wires with an outer sheath or covering of moisture-resistant, nonmetallic material, Figs. 2-4 and 2-5; and FLEXIBLE ARMORED CABLE, Fig. 2-6. Armored cable, commonly referred to as BX, comes in 2 and 3-wire types. Wires with insulation and a bare

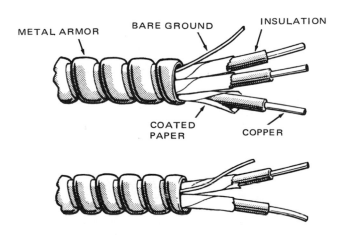

Fig. 2-6. Metallic armored cable. Above. Three-wire cable with bare ground.
Below. Two-wire cable with bare ground.

ground wire are twisted together. This grouping of wires is then wrapped in coated paper and covered with self-locking steel armor. Several types of insulation are listed in Fig. 2-7.

WIRE AND CABLE INSULATIONS

INSULATION	LETTER TYPE
WEATHERPROOF	WP
SLOW-BURNING	SB
SLOW-BURNING WEATHERPROOF	SBW
RUBBER:	
Code Compound	R
Heat-Resistant	RH
	RHH
Moisture-Resistant	RW
Moisture and Heat-Resistant	RH—RW
	RHW
Latex (Regular)	RU
Latex (Moisture-Resistant)	RUW
Latex (Heat-Resistant)	RUH
MINERAL (METAL-SHEATHED)	MI
THERMOPLASTIC COMPOUNDS:	
Thermoplastic	T
Moisture-Resistant Thermoplastic	TW
Moisture and Heat-Resistant Thermoplastic	THW
Thermoplastic and Fibrous Outer Braid	TBS
Thermoplastic and Asbestos	TA
VARNISHED CAMBRIC:	
Standard Black	
Heat-Resisting	V
PAPER:	
Solid Type	
Oil-filled	
Oilostatic	
Untreated	
Treated	
ASBESTOS:	
Nonimpregnated	A and AA
Impregnated	AI and AIA
Asbestos-Varnished-Cambric Outer Asbestos Braid	AVA
Lead Covered	AVL
Cotton Braid Covered	AVB
SILICONE ASBESTOS	SA

Fig. 2-7. Types of wire and cable insulation; National Electric Code designations.

CORDS

Flexible electric cords are commonly grouped and designated as Lamp, Heater, and Power or Service Cords.

Conductors

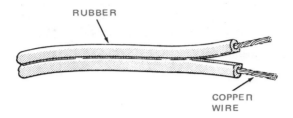

Fig. 2-8. Rubber covered flexible lamp cord.

LAMP OR FIXTURE CORDS: Flexible cords which are used to connect lamps, radios, etc. to outlets. Lamp cord is made up of fine strands to give it flexibility. It is then covered with rubber insulation. In some types of lamp and fixture cords, the insulation is covered with cotton or rayon wrapping. See Figs. 2-8 and 2-9.

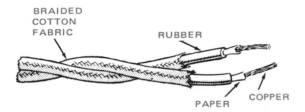

Fig. 2-9. Braided and twisted lamp cord.

APPLIANCE OR HEATER CORDS: For heating appliances, electric irons, toasters, waffle irons, etc., a special cord called a HEATER CORD is used. This has a layer of asbestos wrapped around each rubber-covered wire, and an overall covering of cotton or rayon braid, Fig. 2-10.

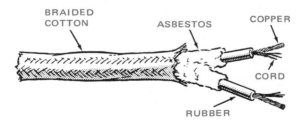

Fig. 2-10. Cord for appliances, heaters.

SERVICE OR POWER CORDS: Electric cords used for large motors, heavy-duty power tools and similar power requirements must be heavy enough to carry the load without overheating. See Fig. 2-11.

House Wiring Simplified

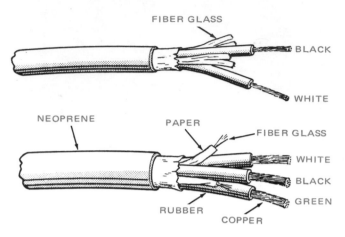

Fig. 2-11. Power or service cord. Above. Two-wire cord. Below. Three-wire cord.

OTHER TYPES AVAILABLE: Fig. 2-12 illustrates some special-purpose cords. As you progress with your house wiring activities, you'll find that many other types of conductors are available.

WIRING MUST MEET CODES

All house wiring must be installed in accordance with City and State Codes or regulations applicable to the work being done, also the National Electric Code. Requirements of the local Utility Company must also be met.

Compliance with Codes usually results in installations free from electrical hazards, but the systems will not necessarily be efficient, convenient, or adequate for good service and future expansion.

Electrical codes are NOT intended as construction guides for untrained people.

The National Electric Code which contains rules and specifications intended to safeguard both people and property from hazards arising from the use of electricity, is sponsored by the National Fire Protection Association, 60 Batterymarch St., Boston, Mass. 02110. The Code book which is revised every three years is available from the Association. A copy of the Code is an important "tool" for everyone interested in house wiring procedures.

USE APPROVED MATERIALS, DEVICES

When doing house wiring, only Underwriters' Laboratories (UL) approved materials and devices should be used. Underwriters' Laboratories which are supported

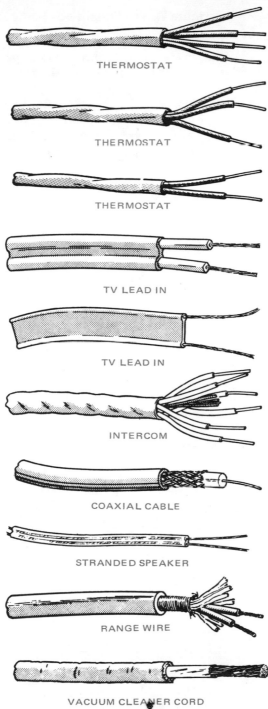

THERMOSTAT

THERMOSTAT

THERMOSTAT

TV LEAD IN

TV LEAD IN

INTERCOM

COAXIAL CABLE

STRANDED SPEAKER

RANGE WIRE

VACUUM CLEANER CORD

Fig. 2-12. Special purpose cords.

by manufacturers, insurance companies, and other interested parties, test materials and devices to see if they meet certain minimum electrical standards.

The Underwriters' stamp of approval (several styles used) which is stamped on or printed and attached to items which have been approved, assures the purchaser that minimum safety requirements as specified by the National Electric Code have been met. The approval stamp, however, is not intended to convey the idea that all approved items are of equal quality.

INSTALLATION MUST BE WORKMANLIKE

All wiring and fixtures should be installed in a neat, workmanlike manner, closely following plans (working drawings) provided for the job.

PROGRESS CHECK — UNIT 2

1. Insulation is a (noncurrent/current) carrying material which insures that the current flow will be through the wire.
2. As the size of the wire decreases, the wire number becomes (larger/smaller).
3. For most house wiring jobs, copper wire numbers _____ and _____ are specified by the building plans.
4. In a two-wire cable, the wire colors are: _____ and _____.
5. In a three-wire cable, the wire colors are:_____ ,_____ , and _____.
6. The neutral or ground wire is always white. True or False?
7. List the two types of cables commonly used in modern house wiring.
   a. _____.
   b. _____.
8. Flexible electric cords are grouped under the following three headings.
   a. _____.
   b. _____.
   c. _____.
9. All house wiring must be installed according to:
   a. City and state codes.
   b. National Electrical Code.
   c. Local utility company requirements.
   d. All of the above.
   e. None of the above.
10. Only_____ _____approved materials and devices should be used when doing house wiring.

Unit 3

CONDUITS, RACEWAYS

Conduit is a type of tubing (metal or plastic) which is used to enclose and protect electrical wiring.

RIGID STEEL CONDUIT

Rigid steel conduit is available both in galvanized and black enameled types. This comes in 10 ft. lengths, with both ends threaded and a coupling screwed on one end, Fig. 3-1. Sizes range from 1/2 to 6 in., Fig. 3-2. Rigid conduit which is bendable is cut and threaded with the same type tools as used for water pipe.

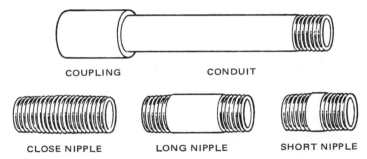

COUPLING   CONDUIT

CLOSE NIPPLE   LONG NIPPLE   SHORT NIPPLE

Fig. 3-1. Rigid steel conduit, coupling, nipples.

SIZE IN.	DIAMETER INCHES EXTERNAL	INTERNAL	THICKNESS IN.
1/2	.840	.622	.109
3/4	1.050	.824	.113
1	1.315	1.049	.133
1 1/4	1.660	1.380	.140
1 1/2	1.900	1.610	.145
2	2.375	2.067	.154
2 1/2	2.875	2.469	.203
3	3.500	3.068	.216
3 1/2	4.000	3.548	.226
4	4.500	4.026	.237
5	5.563	5.047	.258
6	6.625	6.065	.280

Fig. 3-2. Rigid conduit; thickness, internal and external diameters.

19

THINWALL STEEL CONDUIT

Thinwall conduit, which is the type conduit used for most house wiring, is galvanized, light in weight, and is easy to bend and handle. It comes in 10 ft. lengths, without couplings. The wall is so thin it cannot be threaded, so special pressure fittings are used to couple joints together, and to connect the conduit to switch and outlet boxes. See Figs. 3-3, 3-4 and 3-5.

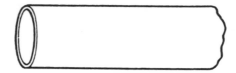

Fig. 3-3. Thinwall conduit.

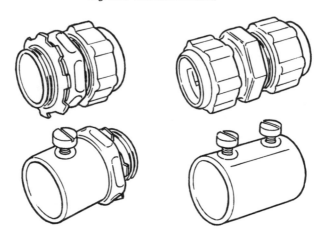

Fig. 3-4. Thinwall fittings (typical).

TRADE SIZE (INCHES)	OUTSIDE DIAMETER (INCHES)	NOMINAL INSIDE DIAMETER (INCHES)	NOMINAL WALL THICKNESS (INCHES)	NOMINAL WEIGHT PER 100 FT. (POUNDS)
3/8	0.577	0.493	0.042	23
1/2	0.706	0.622	0.042	30
3/4	0.922	0.824	0.049	47
1	1.163	1.049	0.057	68
1 1/4	1.510	1.380	0.065	100
1 1/2	1.740	1.610	0.065	114
2	2.197	2.067	0.065	147
2 1/2	2.875	2.731	0.072	230
3	3.500	3.356	0.072	270
4	4.500	4.334	0.083	400

Fig. 3-5. Thinwall conduit; dimensions, weight.

ALUMINUM RIGID CONDUIT

Aluminum rigid conduit comes in 10 ft. lengths. It usually comes threaded on both ends. One end is fitted with a coupling, the other with a plastic protector to protect the threads. See Figs. 3-6 and 3-7.

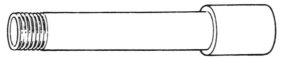

Fig. 3-6. Aluminum rigid conduit.

SIZE IN.	O.D. IN.	I.D. IN.	WALL THICKNESS IN.	WEIGHT PER LENGTH
1/2	.840	.622	.109	3.0
3/4	1.050	.824	.113	4.0
1	1.315	1.049	.133	6.0
1 1/4	1.660	1.380	.140	8.1
1 1/2	1.900	1.610	.145	9.7
2	2.375	2.067	.154	13.2
2 1/2	2.875	2.469	.203	20.8
3	3.500	3.068	.216	27.2
3 1/2	4.000	3.548	.226	32.7
4	4.500	4.026	.237	38.9
5	5.563	5.047	.258	52.9
6	6.625	6.065	.280	78.7

Fig. 3-7. Aluminum rigid conduit; diameter, wall thickness, weight.

FLEXIBLE STEEL CONDUIT

Flexible steel conduit, commonly called "greenfield," is similar in construction to BX, but contains no wires.

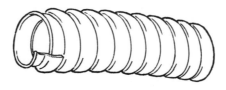

Fig. 3-8. Flexible steel conduit without wires, frequently called "greenfield."

Greenfield is sometimes used with rigid conduit where the runs involve short, difficult bends. Wires are pulled through greenfield after installation. See Figs. 3-8 and 3-9.

SIZE IN.	FEET IN COIL	WT. LBS. PER 100 FT.	SIZE IN.	FEET IN COIL	WT. LBS. PER 100 FT.
5/16	250	15	1 1/4	50	125
3/8	250	25	1 1/2	25	162
1/2	100	47	2	25	213
3/4	50	58	2 1/2	25	263
1	50	102	3	25	313

Fig. 3-9. Greenfield sizes, weights.

RIGID PLASTIC CONDUIT (PVC)

Conduit made from polyvinyl chloride compound (PVC) has established its ability to retain outstanding properties. It will withstand immersion in water or oil; exposure to sunlight, underground moisture and corrosive atmospheres. PVC conduit comes in 1/2 to 4 in. diameters and in 10 and 20 ft. lengths, with plain or threaded ends.

SURFACE RACEWAYS

Metal raceways, Fig. 3-10, are installed on the surface. They provide mechanical protection to conductors while keeping them accessible for wiring changes. On rewiring jobs, the surface installation eliminates dealing with in-wall wiring. The raceways may be finished to match the surrounding surfaces.

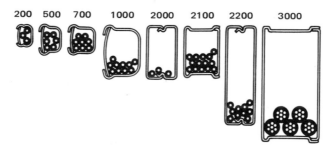

Fig. 3-10. Surface mounted metal raceways.
(Wiremold Co.)

Metal raceways should be selected on the basis of the number and size of conductors to be carried, using specifications and instructions supplied by the manufacturer. Many different systems are available, including raceways in which wires are pulled through, prewired systems, plug-in strips and overfloor systems, Fig. 3-11.

The raceways may be mounted on almost any type of surface using fastening methods shown in Fig. 3-12.

Conduits, Raceways

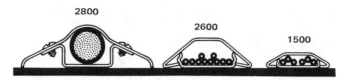

Fig. 3-11. Overfloor type of surface mounted metal raceways.

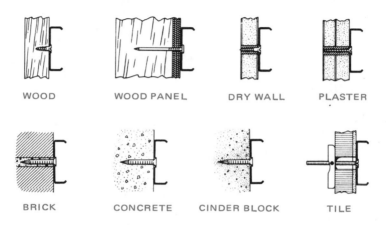

Fig. 3-12. Ways to mount metal raceways.

PROGRESS CHECK — UNIT 3

1. Metal or plastic tubing used to enclose or protect electrical wiring is called _____ .

2. Rigid conduit is cut and threaded with the same tools as used for _____ .

3. Thinwall steel conduit uses special _____ _____ to couple joints and to connect to switch and outlet boxes.

4. Rigid plastic conduit will withstand immersion in_____or _____ .

5. Flexible steel conduit is similar in construction to _____ , but contains no wires.

6. On rewiring jobs, raceway surface installation eliminates dealing with _____ wiring.

7. Metal raceways should be selected on the basis of the _____ and _____ of conductors to be carried.

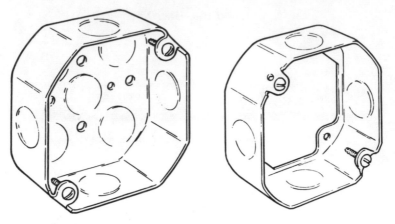

Fig. 4-1. Left. Octagon-shaped junction or ceiling outlet box.
Right. Outlet box extension.

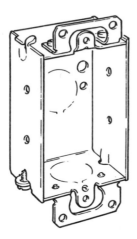

Fig. 4-2. Rectangular-shaped switch or outlet box with square corners.

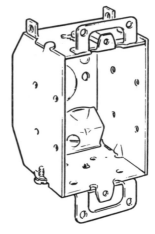

Fig. 4-3. Rectangular-shaped switch or outlet box with beveled corners and clamps for connecting cable.

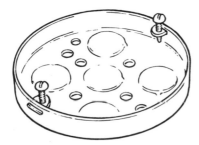

Fig. 4-4. Round, shallow ceiling box. Boxes of this kind are used mostly on "old work" rewiring jobs.

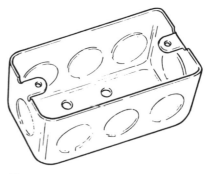

Fig. 4-5. Surface mounted box.

Unit 4

BOXES, COVERS

In modern house wiring, all conductor joints or connections must be housed in approved type boxes, and the boxes must be mounted where they will be accessible for making wiring changes. All switches and outlets must also be housed in boxes, and all fixtures must be mounted on boxes.

BOX CONSTRUCTION

Both steel and nonmetallic boxes are available. The metal boxes are made of heavy galvanized steel, usually 14 ga., and come in four principal shapes, square, octagon, rectangular and circular. See Figs. 4-1 to 4-6 inclusive.

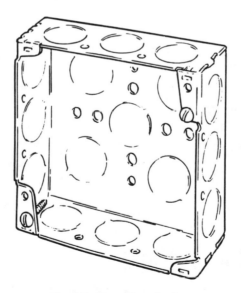

Fig. 4-6. Square junction box.

Outlet boxes made from steel have "knockouts" or machine punched pieces of metal not completely cut loose from the metal box, which are removed to insert conduit and cable. See Fig. 4-7. The knockout with the slot, shown at the left in Fig. 4-7, is removed by inserting a screwdriver and prying. Disk-shaped knockouts can be easily removed, by using a punch or heavy screwdriver to bend the disk outward, and a pair of pliers to break them off.

Fig. 4-7. Left. Two types of box knockouts. Right. Knockout seal.

BOX COVERS OR PLATES

Boxes used for house wiring connections, switches and outlets must be covered. Usually when a fixture is mounted on the box, no other cover is required.

Fig. 4-8 shows an assortment of single, duplex and gang type plates. Television and telephone plates are shown in Fig. 4-9. Most wiring jobs can be completed using one or more of the plates shown. Plates for many specialized purposes are also available.

Fig. 4-8. Single, duplex and gang type box plates or covers.

Boxes, Covers

Fig. 4-8. (Continued) Additional box plates or covers.

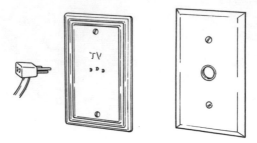

Fig. 4-9. Examples of specialized plates. Left. Television. Right. Telephone.

PROGRESS CHECK — UNIT 4

1. Boxes must house all conductor joints or connections. True or False?
2. Boxes should be mounted so they will be accessible for making wiring changes. True or False?
3. Boxes must house all switches and outlets. True or False?
4. Boxes must be used to mount all fixtures. True or False?
5. List the four basic shapes of metal boxes.

    a. _____.

    b. _____.

    c. _____.

    d. _____.

6. Punched pieces of metal not completely cut loose from the box and removed to insert conduit and cable are called _____ .
7. Usually a cover is not required on a box used for:

    a. Wiring connections.

    b. Switches.

    c. Outlets.

    d. Fixtures.

Unit 5

SWITCHES

SINGLE-POLE SWITCHES

The single-pole switch, Fig. 5-1, with the two terminals, is used to turn on and off, one light or appliance from a single location. Fig. 5-2 shows how a typical single-pole switch works.

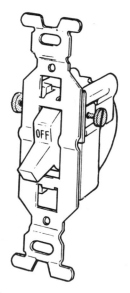

Fig. 5-1. Single-pole switch.

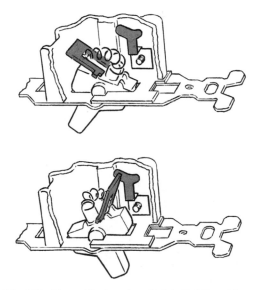

Fig. 5-2. Above. Single-pole switch in "off" position. Below. Same switch in "on" position.

A different (more quiet) type of single-pole switch is illustrated in Figs. 5-3 and 5-4. Fig. 5-5 shows a mercury-operated switch that is completely silent. Electrical contact in this switch is made by mercury moving within a hermetically (airtight) sealed button. There are no mechanical parts to snap or click.

A delayed-action switch is shown in Fig. 5-6. With a delayed-action switch you turn the switch off. As you walk away the light stays on 30 to 60 seconds. Then, the light turns off automatically.

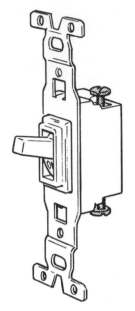

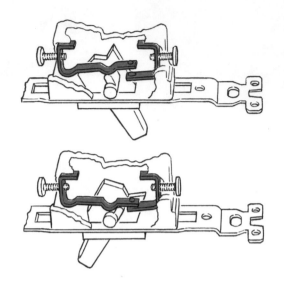

Fig. 5-4. How quiet type switch, shown in Fig. 5-3, works.

Fig. 5-3. Single-pole switch
of quiet type.

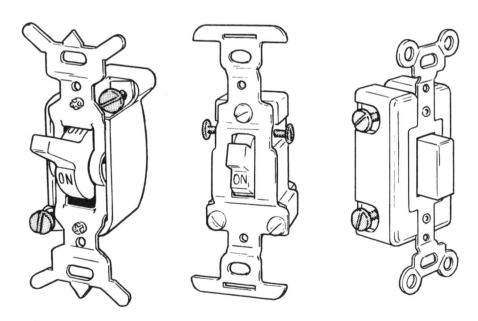

Fig. 5-5. Left. Single-pole mercury operated switch which is completely silent in operation.
Fig. 5-6. Center. Delayed action switch. Fig. 5-7. Right. Single-pole touch switch.

The drawing, Fig. 5-7, shows a single-pole touch switch. The button is pushed to turn the light on and off.

Canopy switches . . . tumbler, pull chain, and push button, Fig. 5-8, are small size, compact switches mounted in the canopies of lighting fixtures, to control the lamps at the fixtures.

A feed through switch inserted in a portable cord is shown in Fig. 5-9.

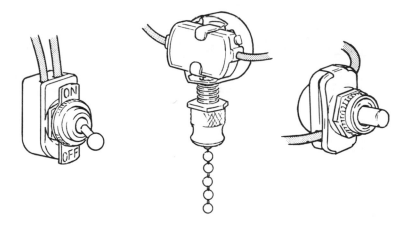

Fig. 5-8. Small canopy switches. Left, Tumbler. Center, Pull chain. Right, Push button.

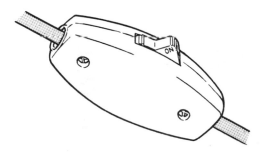

Fig. 5-9. Feed through switch used in portable cord.

THREE-WAY SWITCHES

Three-wire switches, Fig. 5-10, are used to control lights from two locations. Note the three terminals. Two three-way switches are required for each installation. Fig. 5-11 shows how a three-way switch works. See Figs. 12-9, 12-10, and 12-11, pages 95, 96, and 97, for three-way switch wiring diagrams.

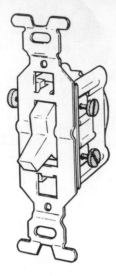

Fig. 5-10. Three-way switch.
Note the three terminals.

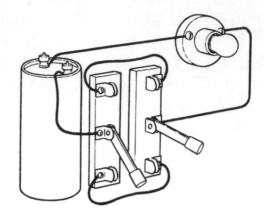

Fig. 5-11. How three-way switch works. The circuit is completed by moving both switches either up or down.

FOUR-WAY SWITCHES

Four-way switches (each switch has four terminals), Fig. 5-12, are used when we want to control an electrical circuit from three points. When using a four-way switch, it is necessary to use two, three-way switches in the circuit, and install a four-way switch between the three-way switches. Fig. 5-13 shows how a four-way switch works. For an installation diagram, see Fig. 12-12, page 98.

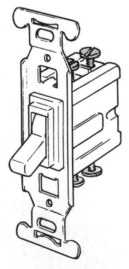

Fig. 5-12. Four-way switch.
Note four terminals.

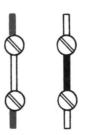
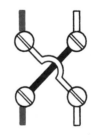

Fig. 5-13. A four-way switch is a special double-pole, double-throw switch, used between two three-way switches to provide an additional switch from which a light may be operated. When the switch is operated, an insulated jumper conducts current to the contact diagonally opposite. Left. First position. Right. Second Position.

Switches

For each additional control point, an additional four-way switch must be inserted between the two three-way switches.

DIMMER SWITCH

The dimmer switch, Fig. 5-14, may, by the turn of a knob, be used to switch a light from bright light (100 percent) to a dim light (approximately 25 percent). Installation is simply a matter of replacing the regular switch with a dimmer switch.

A dimmer switch of the type shown should not be used to control wall outlets, fluorescent lights, appliances or motor-driven equipment.

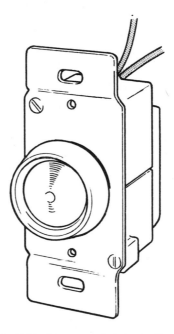

Fig. 5-14. Dimmer switch (typical).

NOTE: Switches illustrated and described in this unit are typical examples of switches commonly used. Numerous other switches are available.

USING INTERCHANGEABLE DEVICES

Interchangeable devices, Fig. 5-15, permit flexibility in an electrical installation. Two or three devices may be assembled in a standard switch or outlet box and mounted onto the same wall plate. The devices include outlets, switches and pilot lights and may be assembled in any combination or order.

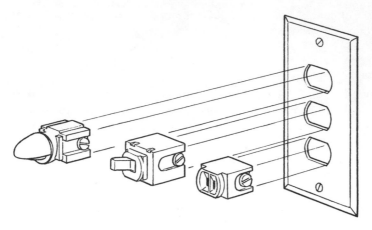

Fig. 5-15. Interchangeable devices.

PROGRESS CHECK — UNIT 5

1. The single-pole switch has _____ terminals and is operated from _____ location.
2. The three-way switch has _____ terminals and is operated from _____ locations.
3. The four-way switch has _____ terminals and is operated from _____ locations.
4. A mercury-operated switch is completely silent because there are no _____ _____ to snap or click.
5. The three kinds of canopy switches are the tumbler, feed through, and the pull chain. True or False?
6. When using a four-way switch, it is necessary to use two _____ switches in the circuit.
7. A dimmer switch should never be used to control:
   a._____ .
   b._____ .
   c._____ .
   d._____ .

Unit 6

OUTLET RECEPTACLES

OUTLETS

Receptacles are used to plug in portable devices such as lamps, toasters, radios, television sets, etc. Numerous types are available.

TYPICAL EXAMPLES

A duplex outlet without a grounding terminal is shown in Fig. 6-1. This type outlet should be used as a replacement for an outlet in an existing electrical system which uses outlets of the same kind, in a system which is not grounded.

Fig. 6-1. Duplex outlet without grounding terminal.

Fig. 6-2 shows a duplex outlet with a green hex head screw terminal for a grounding wire, and a break-off fin. The grounding wire is connected to the green

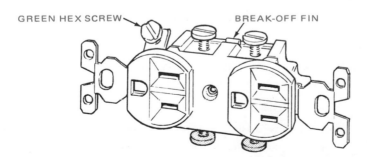

GREEN HEX SCREW   BREAK-OFF FIN

Fig. 6-2. Duplex outlet. Note green hex screw terminal for grounding wire, also break-off fin.

covered or bare wire in cable. When the metal break-off fin is removed (with a screwdriver) there is no connection between the two outlets and the receptacle may be used to provide outlets for two separate circuits.

A single receptacle (240V) with tandem blades and U-shaped ground is shown in Fig. 6-3. The green terminal connects to the bare or green wire in the cable. Receptacles of this type are commonly used for air conditioners and other similar installations.

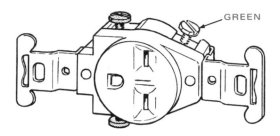

GREEN

Fig. 6-3. Receptacle (240V) for tandem blades and U-shaped ground.

A receptacle for three wires, 240V is illustrated by Fig. 6-4. Receptacles of this kind provide for easy connection and disconnection of equipment such as dryers, ranges, etc.

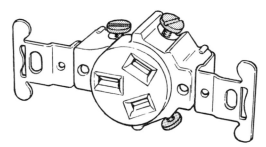

Fig. 6-4. Receptacle for three wires, 240V.

Fig. 6-5, illustrates a single receptacle (240V) for horizontal and vertical blades and U-shaped ground. This is for large air conditioners, heavy power tools, garden equipment, etc. The green terminal connects to the bare (or green covered) wire in the cable.

A no-shock safety duplex outlet is shown in Fig. 6-6. In this receptacle, self-closing outlets prevent small children from getting shocked by inserting metal objects. To use a no-shock type outlet, insert plug, twist to right a quarter turn and push in plug. When the plug is pulled out the rotary cap snaps shut.

Outlet Receptacles

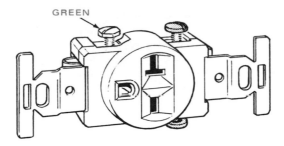

Fig. 6-5. Receptacle (240V) for horizontal and vertical blades and U-shaped ground.

Fig. 6-6. No-shock outlet with self-closing openings.

A weatherproof receptacle for outdoor devices is shown in Fig. 6-7. Usually in house wiring, provision for telephones should be made by installing conduit or cable of approved type which terminates in switch boxes at the locations where telephones

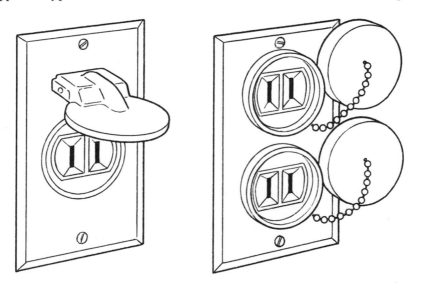

Fig. 6-7. Weatherproof receptacles.

are desired. A special wall plate, left, Fig. 6-8, is provided. The local telephone company should be consulted for complete instructions prior to construction. Fig. 6-8 also shows a special wall plate used for television. TV lead-in cable is run from the antenna to a box at the back of the plate, without the use of conduit.

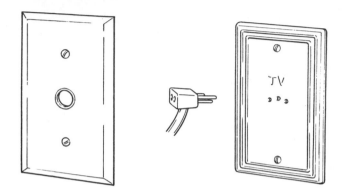

Fig. 6-8. Left. Wall plate for telephone. Right. Plate for TV.

A plastic safety cap for wall outlets which may be used to keep children from inserting dangerous metal objects into outlets, is shown in Fig. 6-9.

Fig. 6-9. Safety cap.

PROGRESS CHECK — UNIT 6

Match the description of the outlet receptacle in the list on the left with the proper use in the list on the right.

a. Duplex outlet without grounding terminal.

b. Tandem blades and U-shaped ground 240V outlet.

c. Three-wire 240V outlet.

d. Horizontal and vertical blades 240V outlet.

e. Self-closing rotary cap outlet.

f. Special wall plate for antenna lead-in cable.

1.___ Dryers, ranges, etc.

2.___ Large air conditioners, heavy tools, garden equipment.

3.___ Replacement in non-grounded system.

4.___ Television signal reception.

5.___ No-shock, prevents children inserting metal objects.

6.___ Air conditioners.

Unit 7

ELECTRICIANS TOOLS AND EQUIPMENT

In this Unit we will discuss tools that may be used to good advantage in house wiring jobs.

PLIERS

Pliers are available with both insulated and uninsulated handles. Insulated handle pliers should be used when working on or near hot wires. The handle insulation should not be considered sufficient protection alone, and other safety precautions must be observed. Several types of pliers are shown in Fig. 7-1.

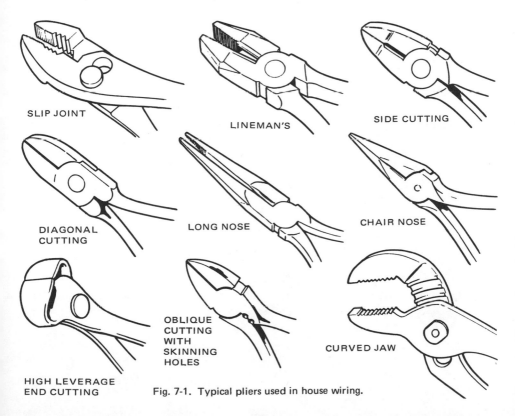

SLIP JOINT

LINEMAN'S

SIDE CUTTING

DIAGONAL CUTTING

LONG NOSE

CHAIR NOSE

HIGH LEVERAGE END CUTTING

OBLIQUE CUTTING WITH SKINNING HOLES

CURVED JAW

Fig. 7-1. Typical pliers used in house wiring.

SCREWDRIVERS

Screwdrivers come in various sizes and with several tip shapes. See Fig. 7-2. Screwdrivers used by electricians should have insulated handles. For safe and efficient use, screwdriver tips should be kept square and sharp. In selecting a screwdriver for a particular job the width of the screwdriver tip should match the width of the screw slot.

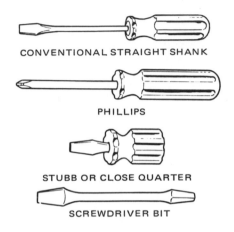

CONVENTIONAL STRAIGHT SHANK

PHILLIPS

STUBB OR CLOSE QUARTER

SCREWDRIVER BIT

Fig. 7-2. Types of screwdrivers used in house wiring.

DRILLING EQUIPMENT

Drilling equipment is needed to make holes in building structures for passage of conduit and wire in both new and old construction. See Figs. 7-3 and 7-4.

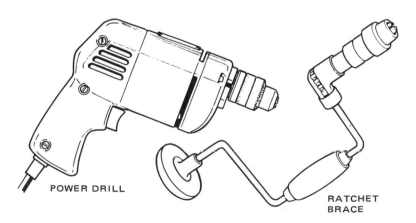

POWER DRILL

RATCHET BRACE

Fig. 7-3. Drilling equipment.

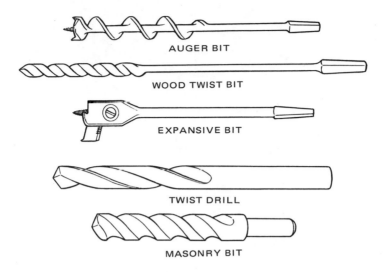

AUGER BIT

WOOD TWIST BIT

EXPANSIVE BIT

TWIST DRILL

MASONRY BIT

Fig. 7-4. Drilling equipment, bits and drills.

SAWING AND CUTTING TOOLS

Saws commonly used by electricians include the crosscut, keyhole, hacksaw, power saw. See Figs. 7-5 and 7-6.

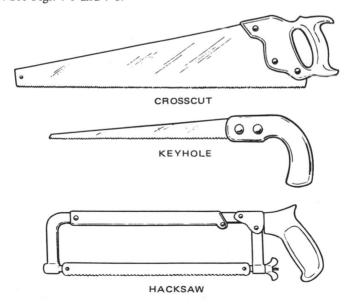

CROSSCUT

KEYHOLE

HACKSAW

Fig. 7-5. Typical hand saws.

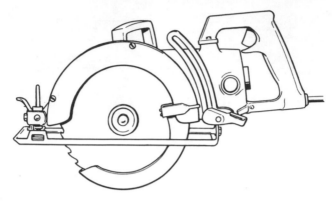

Fig. 7-6. Power saw.

SOLDERING EQUIPMENT

In doing electric wiring, splices and taps (connections made to wire at other than end) should be soldered, unless you use solderless connectors. Typical equipment available for soldering is shown in Fig. 7-7.

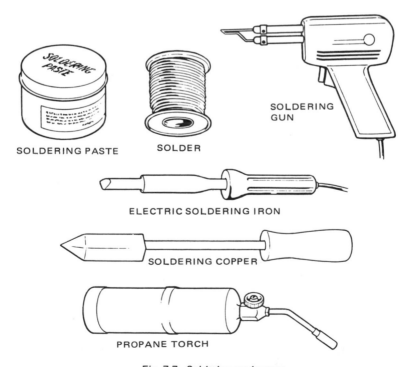

SOLDERING PASTE   SOLDER   SOLDERING GUN

ELECTRIC SOLDERING IRON

SOLDERING COPPER

PROPANE TORCH

Fig. 7-7. Soldering equipment.

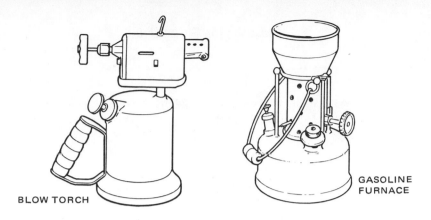

BLOW TORCH

GASOLINE
FURNACE

Fig. 7-7. (Continued) Soldering Equipment.

MULTIPURPOSE TOOL

Fig. 7-8 shows a tool (Stanley) which may be used as a pair of pliers, to cut and strip insulation from wire, to crimp terminals, and to cut screws.

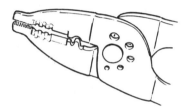

Fig. 7-8. Multipurpose tool.

HAMMERS

Hammers are used with chisels and for nailing and fitting. Fig. 7-9 shows a carpenter's claw hammer, lineman's hammer and a machinist's ball-peen hammer.

Fig. 7-9. Hammers used in house wiring.

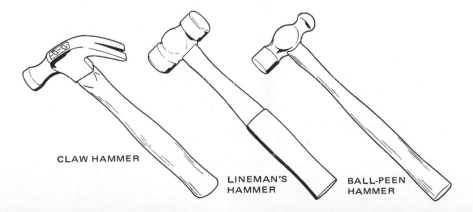

CLAW HAMMER

LINEMAN'S
HAMMER

BALL-PEEN
HAMMER

MEASURING TOOLS

To measure wire length, opening sizes, conduit and other items, the electrician finds considerable use for measuring tools such as the extension rule, push-pull tape rule, and a steel tape, Fig. 7-10.

EXTENSION RULE

PUSH-PULL TAPE RULE

STEEL TAPE

Fig. 7-10. Useful measuring tools.

FISH WIRE OR TAPE

Fish tapes are used to pull (fish) wires through conduits in new work and through wall openings in old work. These tapes are made of tempered spring steel and come in lengths to suit various requirements. See Fig. 7-11.

Wire pulling lubricant or compound (usually a creamy textured compound with wax base) is used to make wire pulling easier.

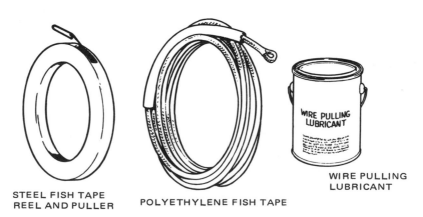

STEEL FISH TAPE
REEL AND PULLER

POLYETHYLENE FISH TAPE

WIRE PULLING
LUBRICANT

Fig. 7-11. Wire pulling equipment.

MISCELLANEOUS TOOLS, EQUIPMENT

Additional tools and equipment needed in handling house wiring jobs includes: conduit bender, pipe cutter, test light, wood chisel, and reamer, illustrated in Fig. 7-12; also unillustrated items such as wrenches, files, fuse pullers, pipe vises, taps and dies for threading conduit, plumb bob for establishing true vertical line, flashlight, test equipment, wire gauge, powder-actuated stud driver.

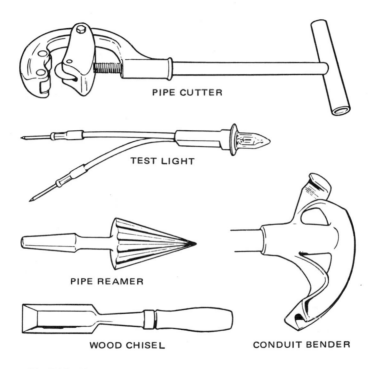

PIPE CUTTER

TEST LIGHT

PIPE REAMER

WOOD CHISEL

CONDUIT BENDER

Fig. 7-12. Miscellaneous tools and equipment used in house wiring.

PROGRESS CHECK — UNIT 7

Identify the tools shown below.

A. _____

G. _____

B. _____

H. _____

C. _____

I. _____

D. _____

J. _____

E. _____

K. _____

F. _____

L. _____

Unit 8

WORKING SAFELY

In house wiring, both the worker and the employer must assume important responsibilities to prevent on-the-job injuries. The employer is responsible for providing proper equipment, equipment maintenance, and safe working conditions. Final responsibility for safe working practices rests with the worker.

SHOCK EFFECTS

If 60-cycle alternating current is passed through a person, from hand to hand or from hand to foot, the usual effects are:

1. At about 1 milliampere (0.001 ampere) the shock may be felt.
2. At about 10 milliamperes (0.010 ampere) the shock may be severe enough to paralyze muscles so a person is unable to release the conductor.
3. At approximately 100 milliamperes (0.100 ampere) the shock may be fatal if it lasts for one second or more.

The amount of shock depends somewhat on conditions at the time and place of contact. If a live wire is contacted while you are standing on a dry wooden floor, the shock may be negligible. If you are standing on a damp floor, your body may become a conductor that leads the current to the ground and a serious shock may result. Shock is far more severe when your hands are damp or wet because water or perspiration reduces the resistance of your body.

When an electric shock is received, current may cause breathing to stop. If the shock is not too severe, breathing may be resumed after a short time, provided a sufficient supply of fresh air is furnished the body through artificial respiration.

FREEING VICTIM

The first person to reach a shocked worker should cut off the current IF THIS CAN BE DONE QUICKLY. If not, the victim should be removed from contact with the energized equipment.

Determine whether the live wire can be removed from the victim, or whether the victim should be pulled away from the wire. Your bare hands MUST NOT be used to pull the victim away. Use a dry board, dry rope, leather belt, coat, overalls or other nonconductor. Be sure to stand on a nonconductor.

Give artificial respiration until a doctor or other competent person familiar with lifesaving methods arrives.

Fig. 8-1. Electricity packs a TERRIFIC wallop. Use your head. Think. Stay alive.

SAFETY SUGGESTIONS

DO NOT TAP INTO LIVE WIRES. Find the switch or circuit breaker, and cut off the current before starting to work on the circuit.

In cutting flexible cable, BX or greenfield with a hacksaw, be sure to hold the cable against a solid object . . . not against your knee.

Remember your eyes are a priceless possession. Protect your eyes. Wear goggles or a face shield where there is a possibility of being injured by flying chips or electric flashes.

Use tools correctly. Make sure all tools are in good working condition. Where there is danger of shock use tools with insulated handles . . . nonmetallic tools where available. Use dry cloth measuring tapes.

Working Safely

Handle and lift objects carefully. When lifting, bend your knees, keep your back as nearly upright as possible.

Periodically run a current leakage check on portable power tools, and check for proper grounding, to make sure they are safe to use.

Check service outlets for polarity and proper grounding.

Keep the floor around your working area clean, dry and free from litter.

Never use a lamp or appliance if the insulation on the cord is worn and ragged. Replace . . . don't splice a broken cord. Discard "beat up" extension cords.

Do not remove a plug from an outlet by jerking on the cord. Pull on the plug.

Before working on electrical apparatus, all rings, wristwatches, bracelets and similar metal items should be removed. Be sure there are no exposed zippers, or metal buttons.

Do not replace a fuse or throw a circuit breaker until the cause of the trouble has been found and corrected.

Burns may result from contact with a hot soldering iron or gun. When soldering, be sure to place the hot iron where it is not likely to be touched by an unsuspecting person.

Many batteries contain acid electrolyte. Drops of the acid can burn your hands and eat holes in your clothing. Battery acid may be neutralized by using a solution of baking soda and water.

On electrical wiring jobs, (on live or near live electrical parts), two men should always work together.

In case of injury, even minor, be sure to get first aid.

When working around electricity, be CAUTIOUS, but not SCARED. Remember that safety and thoughtfulness are closely related.

Remember the ABC's of house wiring . . . Always Be Careful.

PROGRESS CHECK — UNIT 8

1. You can tap into live wires if you use tools with insulated handles. True or False?
2. When lifting, bend your knees, keep your back as nearly upright as possible. True or False?
3. When a lamp cord breaks, it is wise to splice it. True or False?
4. The best way to remove a plug from an outlet is by pulling on the cord. True or False?
5. You should not replace a fuse or throw a circuit breaker until the cause of the trouble has been found and corrected. True or False?
6. When working on electrical apparatus, keep your wristwatch on so you can tell what time it is. True or False?
7. Battery acid may be neutralized by using a solution of baking soda and water. True or False?

Unit 9

SERVICE REQUIREMENTS

Electrical service for today's home should provide for both 120 and 240-volt circuits.

Two power line wires running to a residential service entrance indicate 120-volt service. Three wires indicate 240-volt service is available. One wire will be neutral or ground wire; the other two will be "hot" wires, Fig. 9-1. Lamps ordinarily used in the home, and most plug-in appliances use 120-volt electricity. Many major appliances . . ranges, clothes dryers, etc., need 240 volts.

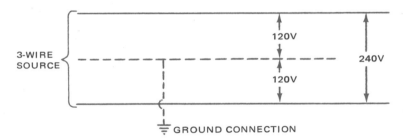

Fig. 9-1. A 240-volt circuit is a combination of two 120-volt circuits.

Service entrance conductors are connected to a main switch, circuit breaker or fuse. The service entrance equipment serves as the junction point from which electricity is dispatched to various parts of the house by a number of smaller wires, called branch circuits.

SERVICE ENTRANCE RATINGS

Service entrance equipment, like appliances and light bulbs, is rated in amperes. It must have sufficient capacity to accommodate the maximum amount of current which will be used at one time . . . both now and in the future.

Today, in most areas, 100-Ampere Service, which may be provided by using No. 2 or No. 3 copper conductors with RHW insulation and a service entrance panel of 100-ampere capacity, is the minimum recommended for new homes.

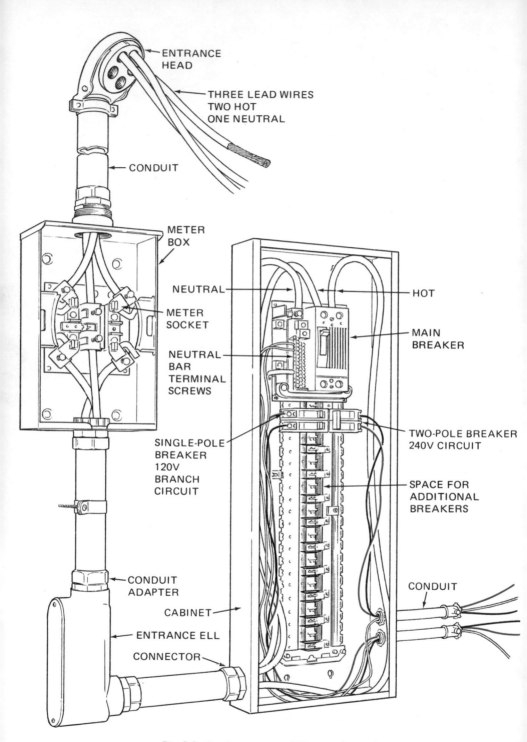

ENTRANCE
HEAD

THREE LEAD WIRES
TWO HOT
ONE NEUTRAL

CONDUIT

METER
BOX

NEUTRAL

HOT

METER
SOCKET

MAIN
BREAKER

NEUTRAL
BAR
TERMINAL
SCREWS

SINGLE-POLE
BREAKER
120V
BRANCH
CIRCUIT

TWO-POLE BREAKER
240V CIRCUIT

SPACE FOR
ADDITIONAL
BREAKERS

CONDUIT
ADAPTER

CONDUIT

CABINET

ENTRANCE ELL

CONNECTOR

Fig. 9-2. Service entrance, 200-ampere capacity.

52

Service Requirements

150, 200-AMPERE SERVICE: Number 1/0 or 3/0 (with type RHW insulation) 3-wire service with a 150 or 200-ampere service panel is desirable where full "housepower" is required, Fig. 9-2. In homes equipped with an electric water heater, range, dryer or central air conditioning, and the usual number of small appliances, 150-ampere service is suggested as a minimum. If the home is to be heated by electricity, or the owner requires more than the normal quantity of appliances, 200-ampere service is usually needed.

To determine the maximum wattage available multiply the amperage rating by the voltage, with a 100-ampere service, multiply 100 x 240 or 24,000 watts. For wattage available with 150-ampere service multiply the amperage rating by the voltage, 150 x 240 or 36,000 watts. A 200-ampere service will make available 200 x 240 or 48,000 watts.

Fig. 9-3 lists a number of appliance wattages. The wattages shown are typical. Actual wattage ratings of different brands of appliances vary considerably. By adding

APPLIANCE	TYPICAL WATTAGE	APPLIANCE	TYPICAL WATTAGE
Air Conditioner (Room)	1200	Hot Plate	1500
Air Conditioner (Central)	5000	Ironer	1650
Attic Fan	400	Lamps, Each Bulb	25-200
Automatic Toaster	1200	Mechanism for Fuel-Fired	
Automatic Washer	700	Heating Plant	800
Broiler	1000	Oil Burner	250
Built-in Ventilating Fan	400	Portable Fan	100
Coffee Maker	1000	Portable Heater	1650
Egg Cooker	600	Radio	100
Deep Fryer	1320	Ranges, Electric	12000
Dehumidifier	350	Refrigerator	200
Dishwasher-Disposer	1500	Rotisserie	1380
Dry Iron or Steam Iron	1000	Roaster	1380
Electric Blanket	200	Sandwich Grill	1320
Electric Clock	2	Saw, Radial	750
Clothes Dryer	9000	TV	350
Freezer	350	Vacuum Cleaner	300
Fluorescent Lights		Ventilating Fan	400
(Each Tube)	15-40	Waffle Iron	1300
Griddle	1000	Waste Disposer	500
Hair Dryer	100	Water Heater	3500
Heat or Sun Lamp	300	Water Pump	700

Fig. 9-3. Appliance wattages (typical).

together wattages of appliances and lamps, which may be used at the same time and appliances which may be added in the future, you can get a good idea of service requirement needs for a particular home.

BRANCH CIRCUITS

Modern house wiring circuits may be divided into three general classes..
GENERAL PURPOSE CIRCUITS, APPLIANCE CIRCUITS, and INDIVIDUAL
CIRCUITS.

GENERAL PURPOSE CIRCUITS should be used for lighting outlets in all rooms
and for convenience outlets in all rooms except the kitchen, dining area and laundry.
General Purpose Circuits should be provided on the basis of one 20 ampere 120V
circuit for not more than each 500 sq. ft. of floor space. Each 20-ampere circuit
wired with No. 12 wire will have a capacity of 2400 watts. (15 amp. circuits with No.
14 wire provide 1750 watts and are usually not recommended.)

Convenience outlets with duplex outlets should be provided along the wall every
12 ft. Outlets on General Purpose Circuits should be divided equally among the
circuits.

SPLIT-CIRCUIT WIRING: A split-circuit wiring arrangement which provides two
separate circuits in each outlet box, is shown in Fig. 12-13, page 99. Such wiring is
desirable where there is need to balance a load of heavy appliances.

APPLIANCE CIRCUITS: Usually, there should be a minimum of two 20-ampere
circuits (using No. 12 wire) in the kitchen and dining area, plus one 20-ampere circuit
in the laundry. In the kitchen counter work area, convenience outlets for the
appliance circuits should be provided at least every 4 feet. If more than one
heavy-duty appliance is to be connected to the circuit, using No. 10 wire is advisable.

INDIVIDUAL CIRCUITS: Wire sizes and types of circuit breakers or fuses
required for circuits serving individual pieces of major electrical equipment, such as
ranges, water heaters, air conditioners, space heaters, etc. will depend on the
amperage rating of the appliances which are to be installed.

Fig. 9-4 shows a typical hookup for an electric range or dryer--a 3-wire, No. 6
cable run from a 50-ampere circuit breaker in the main service panel run to a
heavy-duty wall receptacle. A flexible 3-wire cord or "pig-tail" is connected to the
range or dryer terminals. The other end of the cord has a 3-prong plug which fits into
a receptacle. This setup permits the range or dryer to be easily disconnected. The
metal frame of the appliance (range or dryer) should be grounded to the neutral
terminal.

An Appliance Circuit may be designed to supply both 120 and 240-volt current.
Electric ranges usually operate on 240 volts at high heat and 120 volts at low heat.
Dryers generally require 240 volts for heating and 120 volts for lights and motors.

Service Requirements

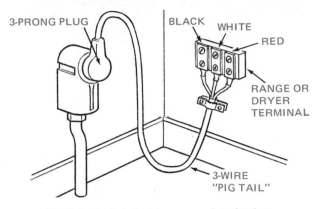

Fig. 9-4. Typical electric range or dryer hookup.

ELECTRICAL SYMBOLS

Electrical symbols, Fig. 9-5, are the electrician's system of "shorthand." They provide a simple way to show on building plans, the electrical service to be provided, and where outlets and switches are to be installed. The use of electrical symbols is shown in Fig. 9-6.

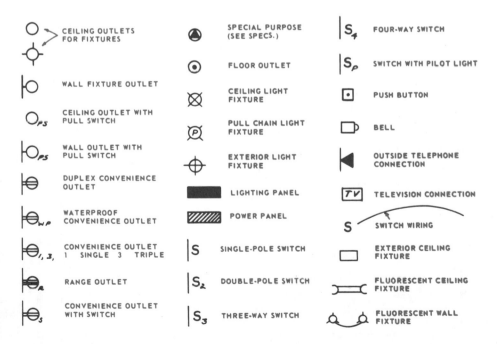

Fig. 9-5. Electrical symbols with which you should become familiar.

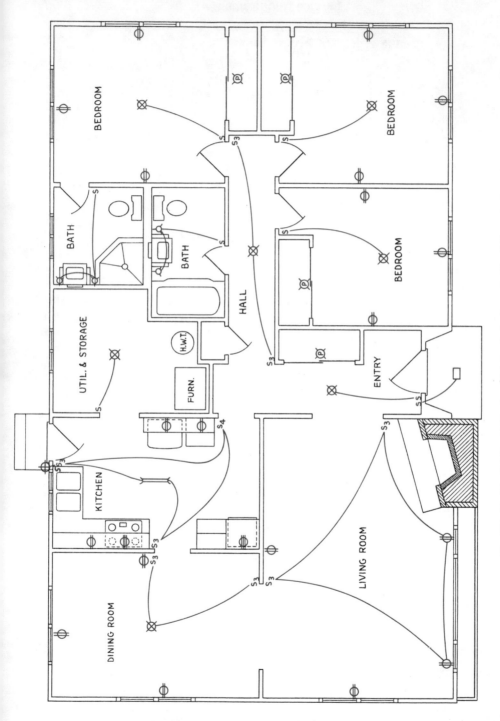

Fig. 9-6. Residential floor plan, in which symbols shown in Fig. 9-6 are used.

BUILDING CODE EXAMPLE

Additional information on the requirements of an adequate wiring system for a modern dwelling may be obtained by studying the Electrical Code of South Holland, Illinois (village where Goodheart-Willcox is located), page 135 of this text.

PROGRESS CHECK — UNIT 9

1. Electrical service for today's modern home should provide for both _____ and _____ volt circuits.
2. A residential service entrance with _____ power line wires indicates 120-volt service.
3. Three-wire 240-volt service uses one neutral or ground wire and two "hot" wires. True or False?
4. Service entrance conductors are connected to:
   a._____ .
   b._____ .
   c. _____ .
5. Today, the minimun recommended service for new homes is _____ amperes.
6. To determine the maximum wattage available, multiply the _____ rating by the voltage.
7. General purpose circuits should be used for lighting outlets in all rooms and for convenience outlets in all rooms except the _____ , _____ , and _____ .
8. Split-circuit wiring provides _____ separate circuits in each outlet box.
9. Wire sizes and types of circuit breakers or fuses required for circuits serving individual pieces of major electrical equipment will depend on the _____ of the appliances to be installed.

Unit 10

BASIC WIRING SYSTEMS

In modern house wiring the basic systems commonly used involve Nonmetallic Cable, Armored Cable, and Metal Conduit.

NONMETALLIC CABLE

Nonmetallic cable is often used on house wiring jobs where the Codes permit. It is easy to install, particularly where it is necessary to "snake" the cable through the walls.

Installation of nonmetallic cable is shown in Fig. 10-1. At the cable ends, strip off the covering allowing at least 8 in. of insulated wire for making connections. Fasten the connector to the outside of the cable cover and insert the cable through the knockout hole of the box. The connector is designed to tightly grip the cable. Screw the locknut on the inside up tightly.

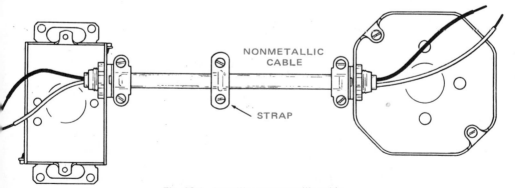

Fig. 10-1. Installing nonmetallic cable.

Run the cable through holes drilled in center of joists or strap the cable every 3 ft. on supporting surfaces such as studs, joists, walls, ceiling. Where the cable runs across joists or through open spaces, support should be provided by running a board (usually 1 x 4) to which the cable is strapped.

In roof areas or attics, run the cable across the top of floor beams, or across face of rafters, and protect by using guard strips.

Strap cable every 4 1/2 ft. (do not use staples). Also, strap within 12 in. of each switch and outlet. In new buildings, straps must be used for all runs whether exposed or concealed. In old work, straps must be used for all exposed runs, but need not be used for concealed runs.

Some Codes required using nonmetallic cable with a ground wire, Fig. 10-2. By using a ground wire, you will have a system that is continuously grounded. This will reduce the danger of shock, if some exposed metal should accidentally become charged with electricity.

BEFORE STARTING ON ANY HOUSE WIRING JOB, CHECK CODES APPLICABLE TO THE JOB. SEE TO IT THAT ALL CODE RESTRICTIONS ARE CAREFULLY OBSERVED.

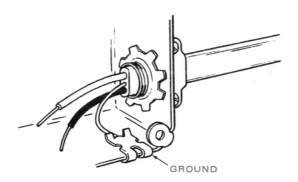

GROUND

Fig. 10-2. Nonmetallic cable. Bare ground wire is being connected to box.

INSTALLING ARMORED CABLE (BX)

Installing armored cable is shown in Figs. 10-3 to 10-7 inclusive. In using a hacksaw to cut the metal armor, Fig. 10-3, place the cable on a solid base, and saw through one section of armor. Twist to break. Use shears to trim off sharp corners. Allow a length of at least 8 in. of insulated wire for making connections in box.

Cutting armored cable with metal-cutting shears is shown in Fig. 10-4. First, bend the armor sharply to buckle it. Then, grip the cable on both sides of buckled point and twist against direction of spiral to open one turn of cable winding. Cut with snips. Trim off sharp corners. Removing armor exposes water-repellent paper wrapping.

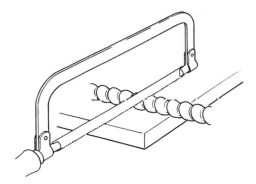

Fig. 10-3. Using hacksaw to cut armored cable.

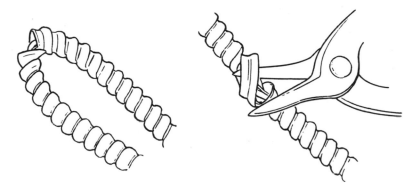

Fig. 10-4. Cutting cable with metal-cutting shears.

Inserting fiber anti-short bushing, Code requirement, between the paper and the wires with bushing turned to cover sharp edges is shown in Fig. 10-5. Remove the paper, slip the connector with the locknut removed over wires and armor. Make sure cable is inserted into connector as far as possible (fiber anti-short bushing must touch

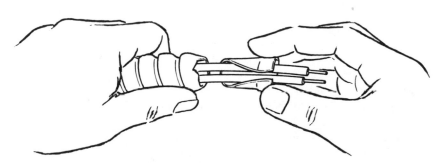

Fig. 10-5. Inserting anti-short fiber bushing between paper and the wires.

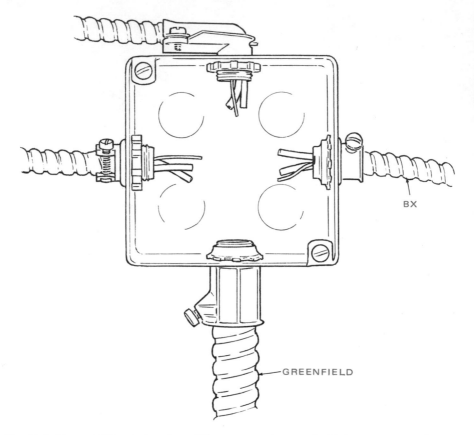

Fig. 10-6. Top and Sides. Using three different types of armored cable box connectors. Bottom. Flexible conduit (greenfield) before wires are installed. In an actual wiring job ends of insulated wires in box would be 8 in. long.

front of connector) then tighten the screw. See Fig. 10-6. In examining the end of the bushing which fits in the box you will notice "peep" holes through which the fiber bushing may be seen. Electrical inspectors use these holes in checking to make sure the bushings have been used.

Some armored cable has a bare ground wire. This should be bent back against armor on the outside and fastened to the screw of the connector, Fig. 10-7. The ground wire, which has less resistance than armored cable, provides a ground connection through the cable from box to box. This is required by some Codes.

Armored cable must be supported by a staple, or strap, every 4 1/2 ft., and within 12 in. of each switch box, except for concealed runs in old work where it is impractical to use straps or staples for support.

IN USING CABLE (NONMETALLIC AND ARMORED) ALL WIRING CONNEC-
TIONS AND SPLICES MUST BE MADE INSIDE METAL BOXES.

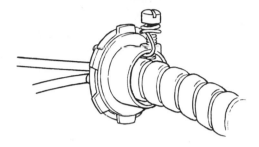

Fig. 10-7. Attaching BX cable grounding wire to screw of connector.

INSTALLING METAL CONDUIT

THINWALL CONDUIT: This provides considerable protection for current-
carrying wires and is required in many localities.

SELECTING CONDUIT: 1/2 in. conduit will carry four No. 14 wires, or three No.
12 wires; 3/4 in. conduit will carry four No. 10 wires, five No. 12 wires, three No. 8
wires.

Thinwall conduit comes in 10 ft. lengths which are joined by threadless
connectors. See Figs. 10-8 and 10-9. Using short couplings between boxes is shown in
Fig. 10-10. Conduit may be cut to length with an ordinary hacksaw using a fine tooth
blade (32 teeth per inch). A reamer or round file should be used to remove sharp
edges and burrs.

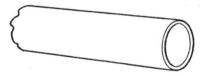

Fig. 10-8. Thinwall conduit shown actual size. Wall thickness of 1/2 in. dia. (outside measurement) is 0.042 in.

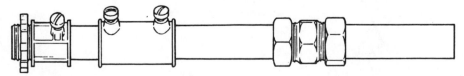

Fig. 10-9. Using thinwall conduit connector and couplings.

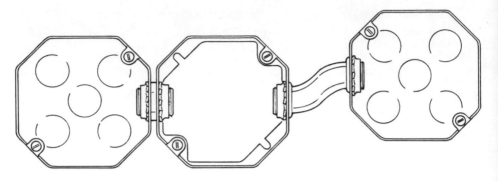

Fig. 10-10. Using thinwall couplings between boxes.

Using a conduit bender to bend thinwall conduit is shown in Fig. 10-11. To make a smooth, even bend, take short bites. Be sure to follow instructions provided by the manufacturer. A little experimenting with a bender will enable you to get the knack of using it. Do not have more than four quarter bends in a run of conduit from one box to another. Avoid short bends.

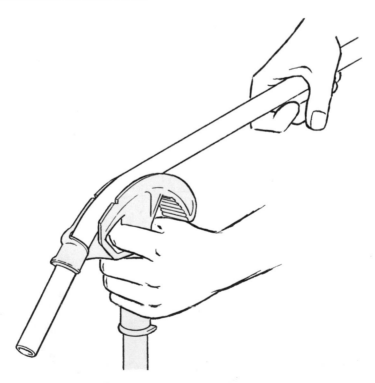

Fig. 10-11. Using conduit bender.

Where conduit runs along the side of a joist or stud, it should be supported every 6 to 8 ft. with a pipe strap or clamp. Where conduit runs horizontally across wall studs, notches should be cut to provide a channel for the conduit, Fig. 10-12.

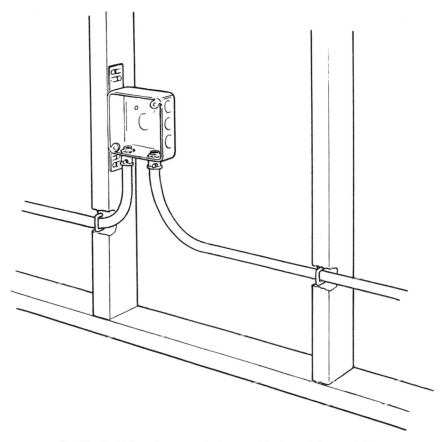

Fig. 10-12. Wall studs are notched to provide channels for conduit.

Notching of studs may be eliminated to a considerable extent by running the conduit across subfloors. When the conduit is in place, furring strips are run up to the conduit. Finish flooring is laid over the furring strips, Fig. 10-13.

Threadless connectors are used to connect the conduit to the metal boxes. Insert conduit through box knockout opening, and tighten the nut on the inside.

In new work the conduit is put in place before the house is completely built. After the walls have been finished, wires are run through the conduit and connected to switches and outlets.

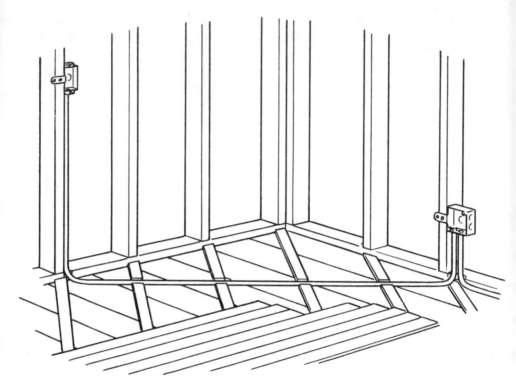

Fig. 10-13. Running conduit across subfloor.

RIGID CONDUIT: Rigid conduit, Fig. 10-14, comes in both black and galvanized types, and looks very much like water pipe. The principal differences are that the conduit is softer, making it easier to bend. It is closely inspected for sharp projections on the inside that might cut the insulation from wires when pulling them through the conduit.

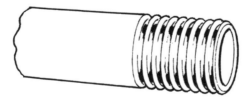

Fig. 10-14. Rigid conduit which is threaded like water pipe.

Rigid conduit comes in the same sizes as water pipe, 1/2 in., (inside measurement) 3/4 in., 1 in., 1 1/4 in., and larger. It may be cut and threaded with the same tools used for water pipe.

PROTECTIVE GROUNDING SYSTEM

The modern house wiring system provides for a protective grounding system, which is independent of the white ground wire. With this setup, should some exposed metal part of tool or appliance become an electrical conductor and be charged with electricity, the charge will be passed harmlessly to the ground.

The grounding system makes use of grounding type outlet receptacles which have the usual two slots for the two blades of the plug, plus an extra round or U-shaped opening for a third prong. See Fig. 10-15. The third prong is attached to a third or grounding wire in the cord which runs to and is connected to the frame of the appliance or tool.

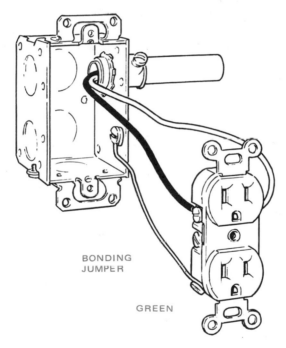

BONDING
JUMPER

GREEN

Fig. 10-15. Connecting grounding wire to box.

In the receptacle the third opening leads to a special terminal (screw head colored green).

Grounding continuity between a properly grounded outlet box using either cable or conduit and the grounding circuit of the receptacle should be established by using a bonding jumper between the box and the grounding terminal of the receptacle, Fig. 10-15.

The jumper is usually not required if the construction and installation of the receptacle is such that direct metal-to-metal contact is established between the receptacle (of approved type) and the box in which it is installed.

PROGRESS CHECK — UNIT 10

1. Before starting any house wiring job, check the _____ applicable to the job.
2. When making connections with nonmetallic cable, strip off the_____ leaving at least 8 in. of insulated wire.
3. When fastening nonmetallic cable to supports do not use staples. True or False?
4. A continuously grounded system will reduce the danger of _____ .
5. Armored cable can be cut using a _____ or _____ .
6. Fiber _____ are used to protect wires from sharp edges on the ends of armored cable.
7. All wiring connections and splices in nonmetallic and armored cable must be made inside metal boxes. True or False?
8. Thinwall conduit may be cut with a _____ and joined using _____ connectors.
9. Using a conduit bender, it is good practice to make up to nine bends in a run of conduit from one box to another. True or False?
10. Rigid conduit may be cut and threaded with the same tools used to cut and thread _____ .

Unit 11

BASIC WIRING PROCEDURES

In 120-volt house wiring there are two wires; a "hot" power carrying wire, and a "neutral" ground wire. See Fig. 11-1.

Fig. 11-1. Two wire, 120-volt circuit consists of one black (hot) wire and one white (neutral) wire.

COLOR CODING OR POLARIZING

Wires throughout the system are identified by color to make sure "hot" or current-carrying wires will be connected to "hot" wires, and the neutral or ground wires run in continuous, uninterrupted circuits.

To maintain the identity of the conductors, the Code (National Electric Code) requires that the neutral or ground wire be white. The hot wire may be black or color other than green. Using green wire in a protective grounding system is described in Unit 10, page 67.

The white wire must be grounded at the main switch (connected to the earth through another conductor) and run to each 120-volt outlet without being interrupted by fuses or switches.

The ground wire from the box to the water main need not be insulated. Minimum size ground wire which may be used is No. 8. Additional information on ground wire size may be obtained from Code table 250-94a, page 145. In connecting the ground to the cold water pipe, a jumper must be installed around the water meter, Fig. 11-2, if the connection is made on the building side of the water meter (side not coming into building from outside).

A white wire is actually a current-carrying conductor for the 120-volt circuit too, even though it is called a neutral or ground wire. The white wire is a very important part of the wiring system. It must be insulated throughout its length, and should be treated with the same respect as the black, or hot wire.

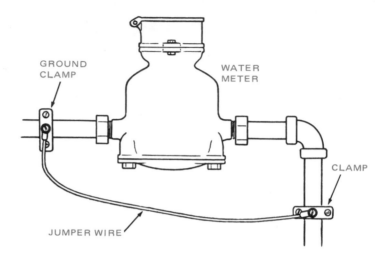

Fig. 11-2. Jumper wire installed around water meter.

THREE-WIRE CIRCUIT

A three-wire circuit includes two hot or colored wires, and one white or neutral wire. Each colored wire provides 120 volts. Connecting the two colored wires, and a white wire to an appliance such as a range or water heater, provides 240 volts, Fig. 11-3.

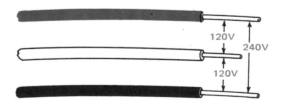

Fig. 11-3. Three-wire circuit which provides 240 volts for appliance use.

CONNECTING WIRES TO TERMINALS

The black or hot wire should be connected to brass-colored terminals on outlet receptacles, switches, fuse and circuit breaker terminals, and to black wires on lighting fixtures. The white or silver-colored wires should be connected to the light or silver-colored terminals on all receptacles, and to the white wires on lighting fixtures. See Fig. 11-4.

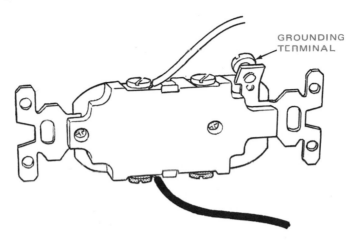

Fig. 11-4. In connecting wires to outlet receptacle, the black or hot wire should be connected to the brass colored terminal and the white (ground) wire to the light or silver colored terminal.

Switches are always connected into the black wire, Fig. 11-5.

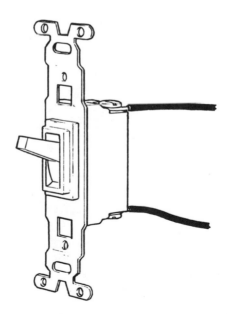

Fig. 11-5. Both terminals of switches are always connected to the black (hot) wire.

In making a connection to a switch or receptacle terminal, remove about 7/8 in. of the insulation, (using wire stripping tool is easiest way) and connect with wire bent to form a clockwise loop, Fig. 11-6. If insulation is removed with a knife, taper insulation at end, so wire will not be nicked by the knife, Fig. 11-7.

Fig. 11-6. Left. Wire is bent clockwise to form loop to make connection to terminals. Fig. 11-7. Right. In using a knife to remove insulation, taper the insulation. This prevents nicking the wire with the knife and weakening it.

USING SOLDERLESS CONNECTORS

In modern house wiring most soldering is eliminated by the use of solderless connectors, Fig. 11-8. The connectors are made of insulating material, so no taping is required. Procedure: Just screw connector over wires, as shown, being sure no bare copper wire is exposed. Remember: All solderless connections must be enclosed in metal boxes.

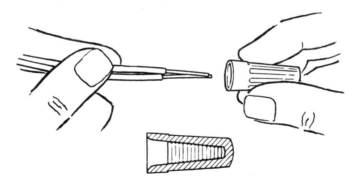

Fig. 11-8. Using solderless connector.

CONNECTIONS REQUIRING SOLDERING

In cases where solderless connectors are not used, all wire splices and taps must be soldered. In splicing two wires, Fig. 11-9, remove about 3 in. of insulation from each wire. Be sure wires are bright and clean. Cross wires about 1 in. from insulation, then make 6 to 8 turns using your fingers and a pair of pliers. The connection must be tight and securely soldered, using a non-acid flux or rosin-core solder. Heat should be applied to the wire joint so heat from the wire melts the solder. Cover with plastic tape which does the work of both rubber and friction tapes. Be sure to provide insulation equal to original wire insulation.

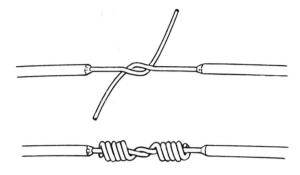

Fig. 11-9. Splicing two wires.

A tap splice (connecting end of wire at point on continuous wire) is shown in Fig. 11-10. Clean wires, wrap loose end around continuous wire. Solder, tape.

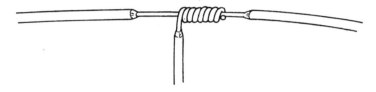

Fig. 11-10. Tap splice.

Fig. 11-11 shows a "pigtail" splice which is sometimes used in outlet boxes to attach fixture leads, or on other connections where there is no pull on the wires.

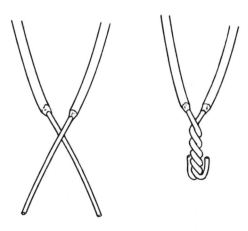

Fig. 11-11. Pigtail splice.

Using a multipurpose tool to install a solderless terminal is shown in Fig. 11-12. Plastic insulation is permanently bonded to the terminal.

Fig. 11-12. Installing solderless terminal on heavy wire.

INSTALLING BOXES

All wiring connections (wire ends and splices where insulation has been removed) and all switches and outlets must be enclosed in approved type boxes. The boxes should be so they are accessible without damaging wall framing or covering.

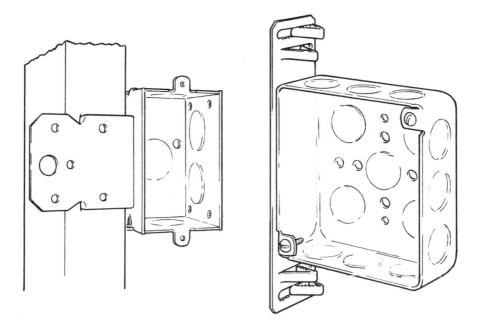

Fig. 11-13. Left. Switch box with side mounting bracket. Nails are used to hold the box in place. Right. Square switch and receptacle box with side mounting bracket.

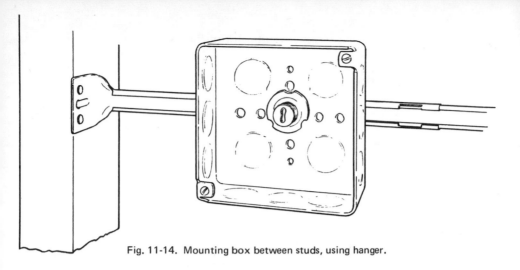

Fig. 11-14. Mounting box between studs, using hanger.

Fig. 11-13 shows switch and outlet receptacle boxes with side mounting brackets. Fig. 11-14 shows how an outlet box may be mounted between wall studs using a metal hanger. Remove the center knockout (machine punched circles which are not completely severed) in the bottom of the box and slip the box over the fixture stud (use screwdriver to remove nut). Tighten nut on inside of box.

The box should be mounted so the front edge is not more than 1/4 in. below the finished surface of the wall or ceiling. If the wall is of combustible material the front edge must extend out flush with the finished surface. Various covers are available for use with square boxes, which will raise the front edge flush with the surface, Fig. 11-15. See also Fig. 4-8, page 26.

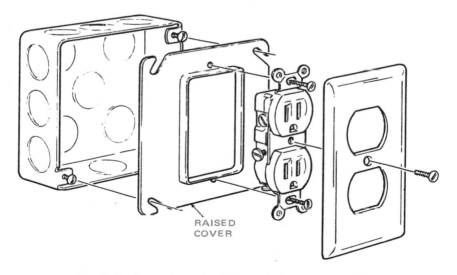

RAISED
COVER

Fig. 11-15. Square boxes should be used where extra space is needed for wires. Various types of raised covers are available.

MOUNTING FIXTURES

Fig. 11-16 shows a porcelain lamp holder mounted directly onto a standard junction box.

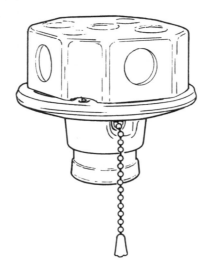

Fig. 11-16. Porcelain lampholder mounted on junction box.

In Fig. 11-17, note the use of a hickey (reducer) which screws onto the threaded fixture stud, and takes a threaded nipple. When light fixture is installed the nipple extends through the fixture which is held in place by a cap screwed onto the nipple.

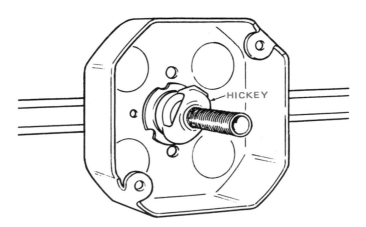

Fig. 11-17. Hickey screwed onto threaded fixture stud takes threaded nipple which supports fixture.

Providing fixture straps and threaded nipples for fixture support is shown in Fig. 11-18.

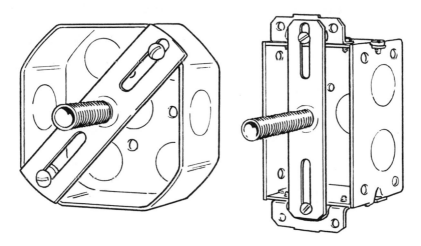

Fig. 11-18. Fixture straps mounted on boxes take threaded nipples which support fixtures.

Provide box entry points for conduit and cable by removing knockouts. Use "beat-up" screwdriver and hammer to tip circle; and pliers to twist it off. Be careful not to remove more knockouts than necessary. Make connection to box, as shown in Fig. 11-19.

GANGING METAL BOXES: Some metal switch boxes are made so that they may be put together or "ganged" to provide additional space. Simply remove one side of each box, fit boxes together and tighten screws.

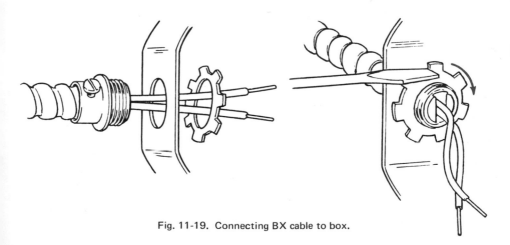

Fig. 11-19. Connecting BX cable to box.

PULLING WIRE INTO CONDUIT

In new construction, wires should not be drawn through the conduit until the plastering has been finished. Make sure that wires conform to the standard color code. Use one black wire and one white wire in a 2-wire circuit; use one black, one white, and one red wire in a 3-wire circuit.

On short runs wires can usually be pushed through the conduit without using a "fish tape." Where the run is long and several wires are to be inserted in the conduit, a fish tape will be needed, Fig. 11-20. Fish tape is made of stiff, flexible steel . . usually about 1/8 in. wide.

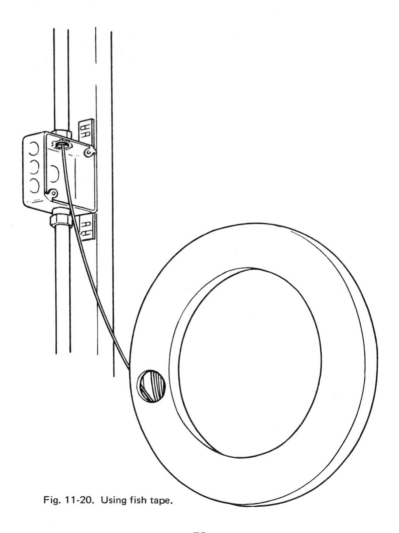

Fig. 11-20. Using fish tape.

Form a small loop on the end of the tape, heat end of steel tape in open flame using blow torch until red to remove temper, then bend to shape as shown in Fig. 11-21. The loop will permit the tape to go around bends as it is pushed into the conduit.

Fig. 11-21. End of fish tape showing loop and how wires are fastened to tape.

In using a fish tape, insert the end of the tape in the conduit. When the end emerges, attach wires to be pulled into the conduit. Be sure there are no sharp ends which might catch at conduit joints. Pull wires through conduit. Using wire-pulling lubricant will help make pulling of the wires easier.

One type of wire lubricant comes in an aerosol can. The lubricant is dispensed into end of conduit as a creamy foam. The lubricant clings to the wires as they are pulled into the conduit. Wire-pulling lubricants also come in liquid and paste forms.

CIRCUIT BREAKERS AND FUSES

Circuit breakers and fuses act as "safety valves" for the wiring system. They protect the wiring from overloads and short circuits.

CIRCUIT BREAKERS

A single-pole circuit breaker of thermal magnetic type is shown in Fig. 11-22. This has a bimetallic element (element consists of two strips of dissimilar metal bonded together) which responds to changes of temperature within the circuit. When excessive current is flowing in the circuit, heat created by the resistance of the bimetallic unit expands each metal at a different rate, causing the strip to bend and open the circuit. Circuit breakers, like fuses, are rated in amperes. They will carry their rated loads continuously, and overload for short periods of time as required to start shop motors, air conditioners, clothes dryers, etc.

After the cause of the trouble has been located and corrected, you simply flip the circuit breaker switch to restore power. See Fig. 11-23.

The drawing, Fig. 11-24, shows the breaker arrangement for typical entrance panel of 100 amp. capacity (large enough for cottage or small home with minimum electrical requirements).

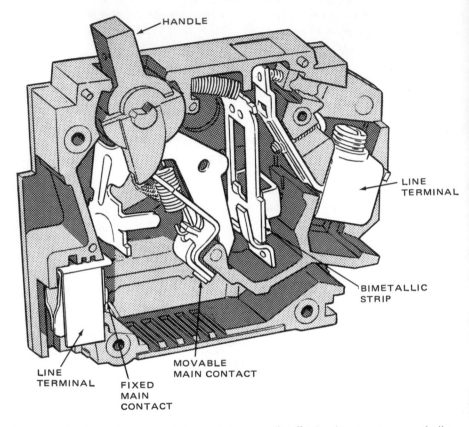

HANDLE

LINE
TERMINAL

BIMETALLIC
STRIP

LINE
TERMINAL

FIXED
MAIN
CONTACT

MOVABLE
MAIN CONTACT

Fig. 11-22. A circuit breaker is a switch in the black or "hot" wire that opens automatically when a predetermined current overload flows through it. Individual breakers are grouped and placed in an entrance panel box. Each individual circuit requires a separate breaker.
(General Electric)

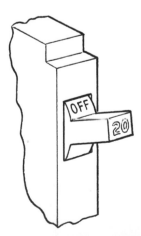

Fig. 11-23. Circuit breaker switch. When overload or short occurs the breaker will switch off automatically. After trouble has been located and corrected, switch is flipped to restore power.

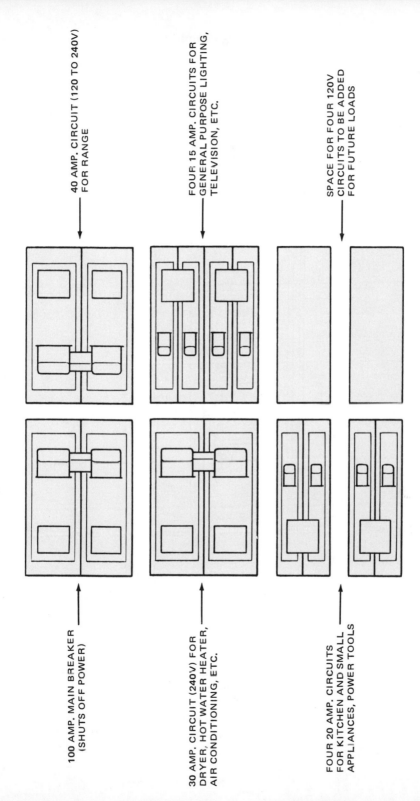

40 AMP. CIRCUIT (120 TO 240V) FOR RANGE

FOUR 15 AMP. CIRCUITS FOR GENERAL PURPOSE LIGHTING, TELEVISION, ETC.

SPACE FOR FOUR 120V CIRCUITS TO BE ADDED FOR FUTURE LOADS

100 AMP. MAIN BREAKER (SHUTS OFF POWER)

30 AMP. CIRCUIT (240V) FOR DRYER, HOT WATER HEATER, AIR CONDITIONING, ETC.

FOUR 20 AMP. CIRCUITS FOR KITCHEN AND SMALL APPLIANCES, POWER TOOLS

Fig. 11-24. Breaker arrangement for 100 amp. entrance panel.

FUSES

An entrance panel which makes use of fuses instead of circuit breakers to protect against shorts and overloads of the circuits is shown in Fig. 11-25. Fuses screw in and out like light bulbs.

MAIN FUSE
PULLS OUT

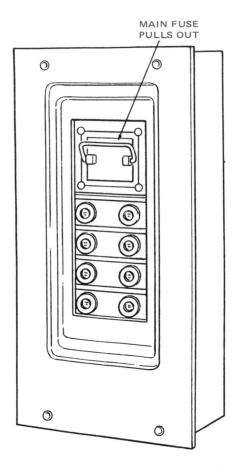

Fig. 11-25. Entrance panel . . . 100 amps., 240 volts, with 8 branch circuits.

HOW PLUG-TYPE FUSE WORKS: In a standard plug-type fuse, as shown in Fig. 11-26, the current passes through the metal strip running across the face of the fuse.

When the fuse is blown by overloading, the metal strip overheats and melts at the weakest point. This breaks the flow of current. When a fuse is blown because of overloading, the fuse window remains clear. If the fuse is blown because of a short

Fig. 11-26. Plug type fuse. The glass top helps prevent shocks when changing fuses.

circuit, the metal strip is heated to a high temperature and vaporizes. This discolors the fuse window. See Fig. 11-27. Screw-type fuses come in capacities up to 30 amperes. Where No. 14 wire is used, 15 amp. fuses are the largest that may be used; 20 amp. fuses are the largest that may be used with No. 12 wire. Cartridge type fuses, Fig. 11-28, as used in house wiring come in capacities up to 100 amperes.

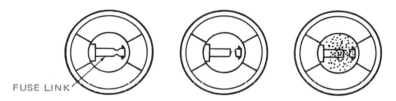

FUSE LINK

Fig. 11-27. Plug type fuses. Left. New fuse. Current passes through strip of thin metal (fuse link). Center. Fuse blown by overload. Right. Fuse blown by short circuit.

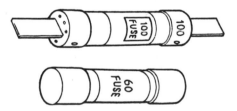

Fig. 11-28. Cartridge type fuses.

Fustats, Figs. 11-29 and 11-30 are tamper-resistant protective devices which provide delayed fuse action. When a short circuit develops, the fuse link blows the same as in a regular fuse. For moderate overloads, instead of the fuse link blowing, the solder cap starts to heat. If the overload continues, the solder in the solder cup softens, and the spring pulls the fuse line out of the solder cup, opening the circuit.

With each different size of fustat (15 amps., 20 amps., etc.) a different fustat adapter is required. The adapter is constructed so when it is screwed into the socket

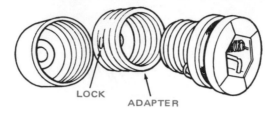

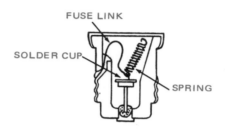

Fig. 11-29. Fustat and adapter.

Fig. 11-30. How fustat is constructed.

it locks in place, and cannot readily be removed. This prevents inserting fuses that are too large.

CIRCUIT BREAKERS THAT SCREW INTO FUSE SOCKETS: Also available for protecting electrical wiring circuit, are miniature circuit breakers that screw into ordinary fuse sockets, Fig. 11-31. The protector trips on dangerous overloads or short circuits. Service is restored by pressing a small button that protrudes from the top.

Fig. 11-31. Small circuit breaker that screws into fuse socket.

CAUTION: WHEN A FUSE IS BLOWN, OPEN THE MAIN SWITCH OR PULL OUT THE SECTION OF THE PANEL LABELED "MAIN" TO CUT OFF THE CURRENT. CORRECT THE CONDITION WHICH CAUSED THE OVERLOAD, THEN REPLACE THE BLOWN FUSE WITH A NEW ONE OF THE PROPER SIZE. MAKE SURE YOUR HANDS ARE DRY, AND STAND ON A DRY BOARD. CLOSE THE MAIN SWITCH OR REPLACE THE PULL-OUT SECTION OF THE PANEL TO RESTORE SERVICE.

ADDITIONAL INFORMATION

The reference section of this book contains a number of useful tables quoted from the National Electrical Code. See pages 139 to 164.

PROGRESS CHECK — UNIT 11

1. To identify the conductors in a house wiring system, the wires are _____ coded.
2. The National Electrical Code requires that the neutral or ground wire be white and the hot wire be some color, like black, red, green, or blue. True or False?
3. A white wire is actually a current-carrying _____ for the 120-volt circuit.
4. Connecting the two colored wires and the white wire in a three-wire circuit provides _____ volts.
5. When connecting wires to terminals, black wires should be connected to _____ colored terminals on outlet receptacles and the white wires should be connected to the _____ colored terminals.
6. Switches are always connected to one white wire and one black wire. True or False?
7. All solderless connections must be enclosed in _____ _____ .
8. Connecting one end of a wire to a point on a continuous wire is called a _____ splice.
9. Wire-pulling lubricant comes in the form of:
    a. Creamy foam.
    b. Liquid.
    c. Paste.
    d. All of the above.
10. Circuit breakers and fuses protect wiring systems from _____ and _____ circuits.
11. When a fuse blows because of a _____ the fuse window remains clear.
12. After a fuse has blown and the trouble corrected, replace the blown fuse with one of the _____ size.

House Wiring Simplified

Unit 12

HOUSE WIRING CIRCUITS

In this Unit, easy-to-understand drawings show how to install wiring required for switches, outlets and fixtures.

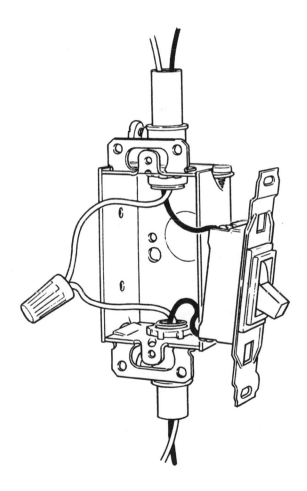

Fig. 12-1. To add switch, cut black wire, and attach both ends to terminal screws of switch. Run conduit ends into new box. Use solderless connector to connect two white wires.

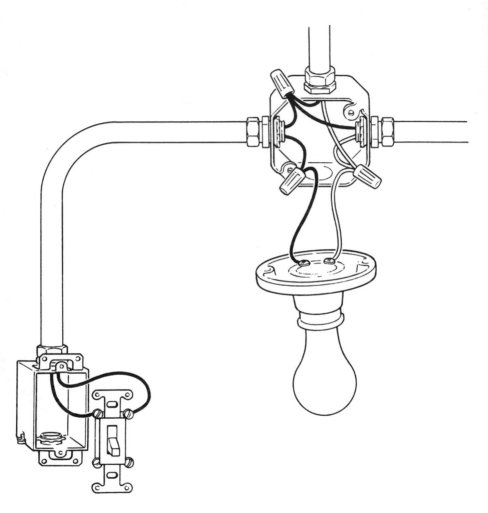

Fig. 12-2. Using wall switch to control light away from switch. Note that all wiring to switch is black wire. If nonmetallic or armored cable is used in wiring instead of conduit, a two-wire cable (one white wire, one black wire) may be used. This special use of a white wire as a hot wire is approved by the Code. Both ends of the white wire should be painted black.

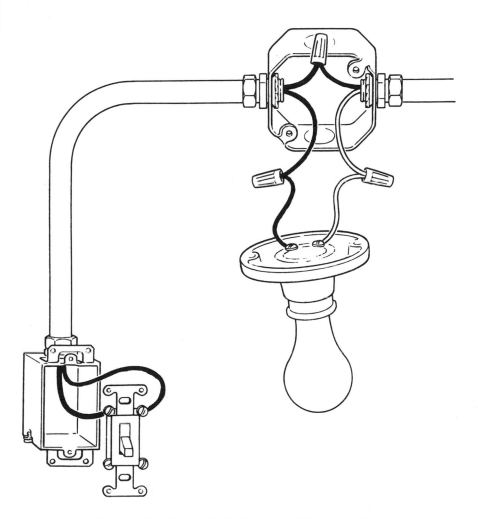

Fig. 12-3. Installing wall switch to control light at end of run.

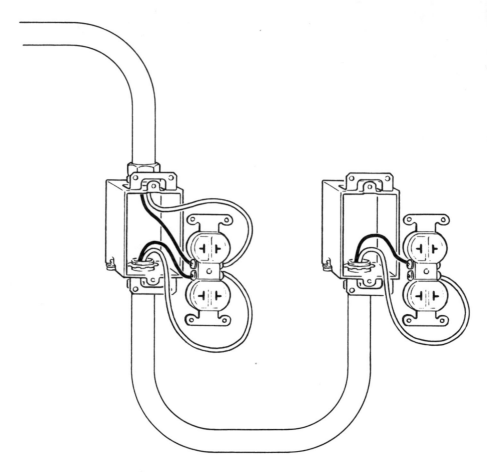

Fig. 12-4. Installing two convenience outlets.

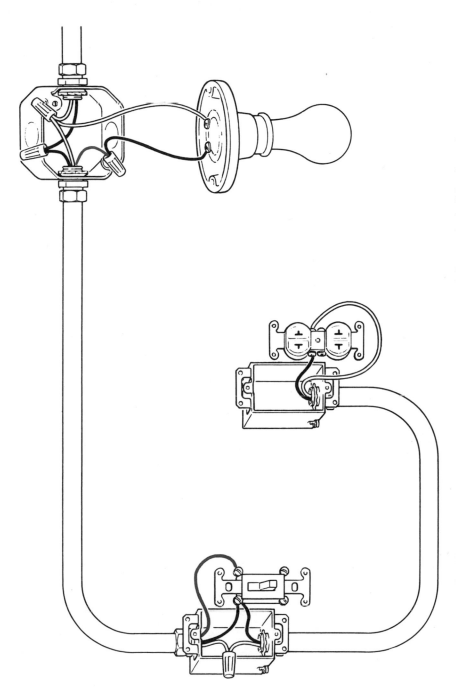

Fig. 12-5. Installing the outlet and switch beyond light. The switch controls the light; the outlet is always hot.

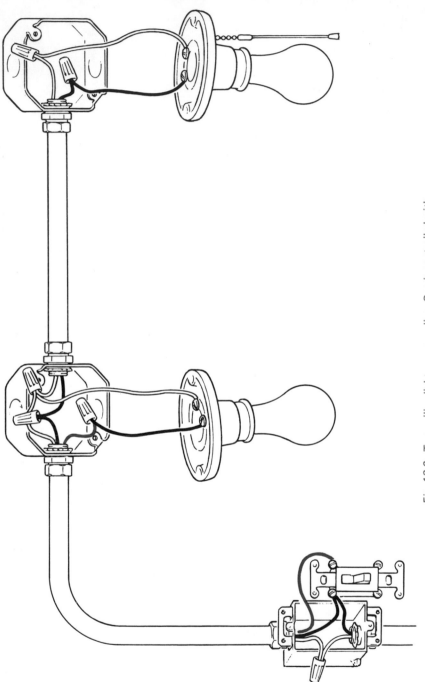

Fig. 12-6. Two ceiling lights on same line. One is controlled with wall switch, the other with pull chain switch.

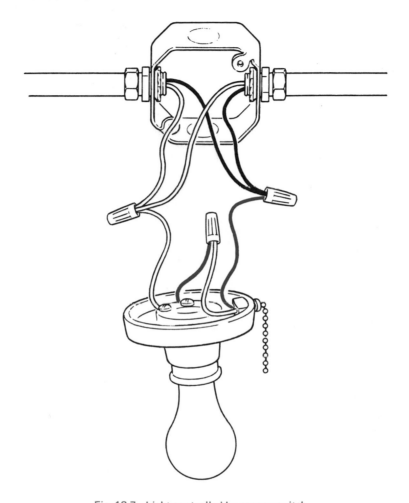

Fig. 12-7. Light controlled by canopy switch.

House Wiring Simplified

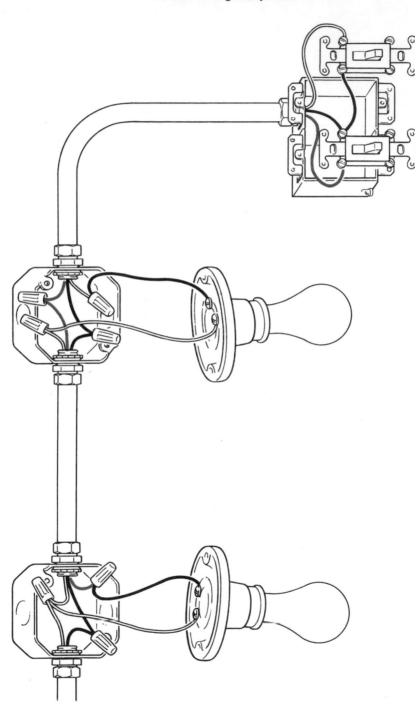

Fig. 12-8. Two ceiling lights operated by individual switches.

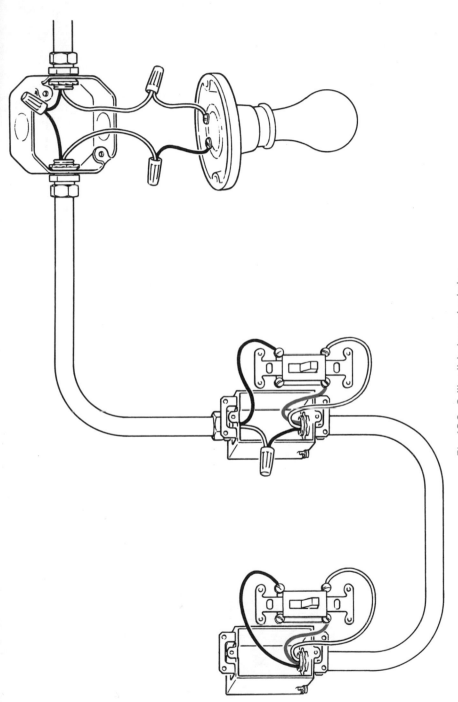

Fig. 12-9. Ceiling ligh: beyond switches, controlled with two 3-way switches.

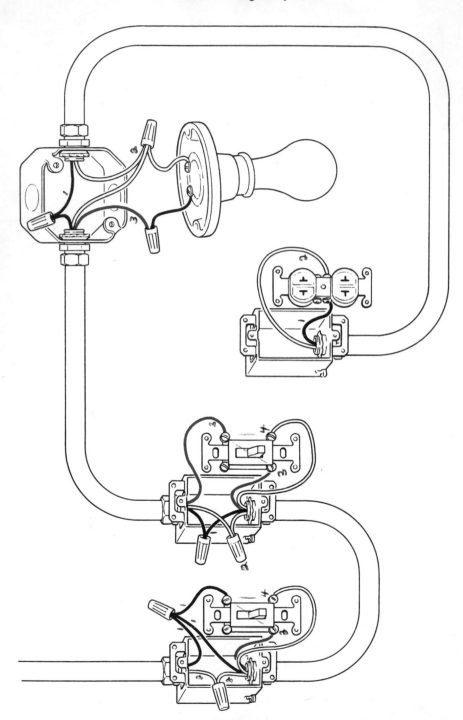

Fig. 12-10. Ceiling light beyond three-way switches, the receptacle is always hot.

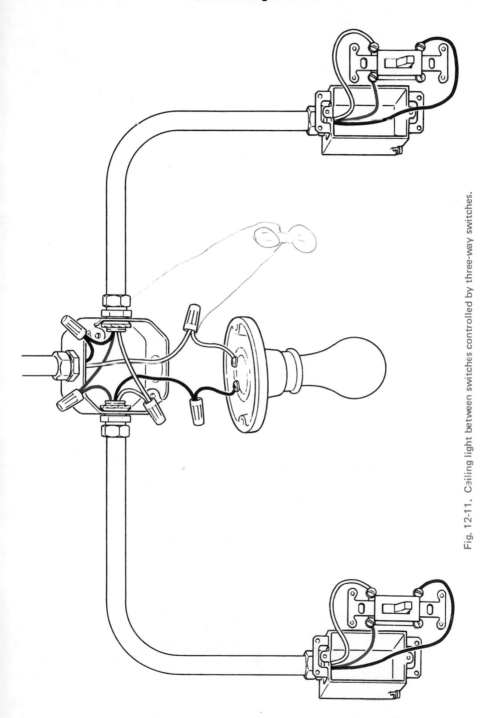

Fig. 12-11. Ceiling light between switches controlled by three-way switches.

House Wiring Simplified

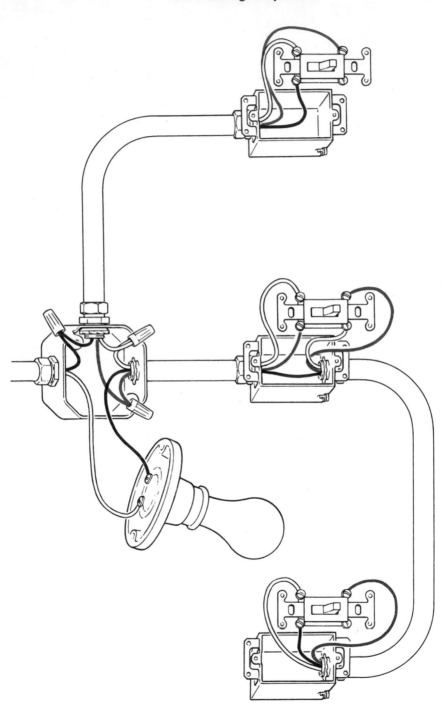

Fig. 12-12. Installing 4-way switch with two 3-way switches to control light from three points.

House Wiring Circuits

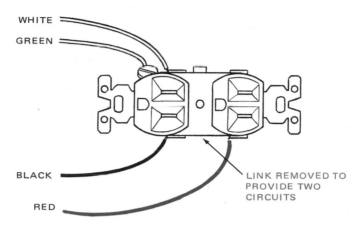

WHITE

GREEN

BLACK

RED

LINK REMOVED TO
PROVIDE TWO
CIRCUITS

Fig. 12-13. Split circuit wiring which provides two separate circuits at each outlet receptacle. Desirable where there are several heavy appliances . . . as in kitchen or workshop. Requires three wires from main switch box (red, black, white) and removal of break-off link (by twisting with pliers) on hot wire side (brass-colored terminal side). The third prong on the attachment plug serving grounded electrical equipment contacts ground through receptacle mounting strap. Shock protection exists only when the ground is completed through a separate grounding wire (bare copper or green insulated) connected to the green terminal, or when the receptacle box is grounded by the third wire or through conduit or armored cable. Either or both circuits may be switch controlled, if desired.

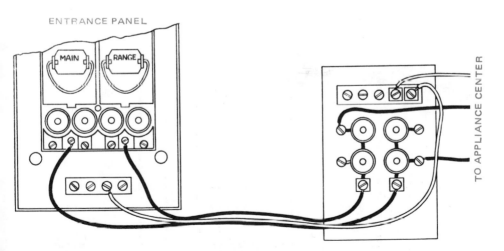

ENTRANCE PANEL

MAIN

RANGE

TO APPLIANCE CENTER

Fig. 12-14. Installing fuse or circuit breaker panel to provide new circuits, when there are no unused circuits on entrance panel. Connect two black wires to power takeoff lugs and white wire to neutral strip of service entrance panel. This provides 120 volts between black and white wires and 240 volts between two black wires.

House Wiring Simplified

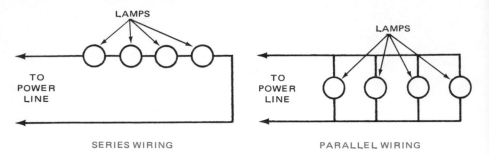

SERIES WIRING   PARALLEL WIRING

Fig. 12-15. Series and parallel circuits. In series circuit electricity to operate first lamp must pass through first lamp, etc. Series wiring is sometimes used when it is desirable to operate several low-voltage lamps (like Christmas tree lights) on line of higher voltage.

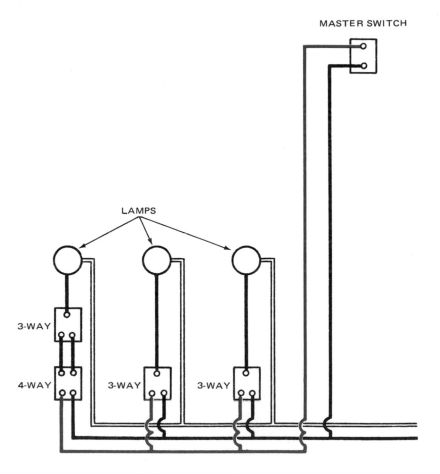

Fig. 12-16. Wiring controlled by master switch.

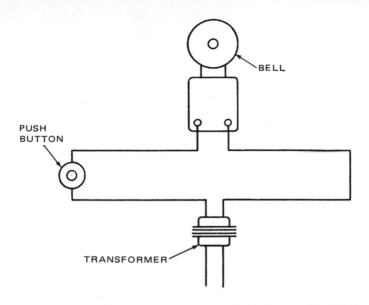

Fig. 12-17. A simple, transformer operated signal system. No. 18 bell wire is ordinarily used to make the hookup from the transformer to the bell.

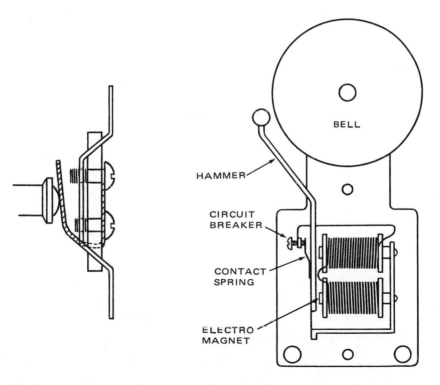

Fig. 12-18. Left. Push-button switch. Pressure against button causes movable contact to meet fixed contact and close circuit. Right. Door bell with principal parts identified.

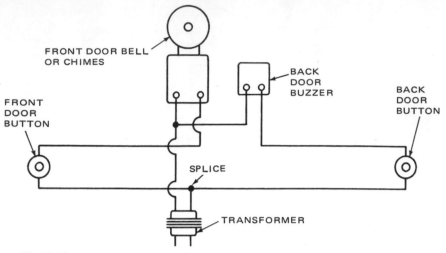

Fig. 12-19. Wiring diagram for front and back entrances, using both bell and buzzer.

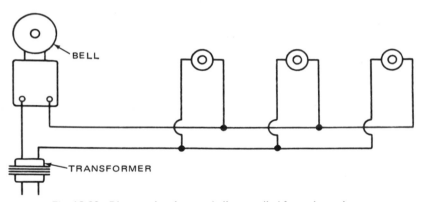

Fig. 12-20. Diagram showing one bell controlled from three places.

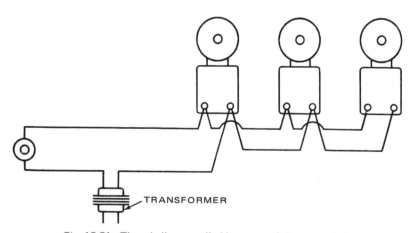

Fig. 12-21. Three bells controlled by one push-button switch.

House Wiring Circuits

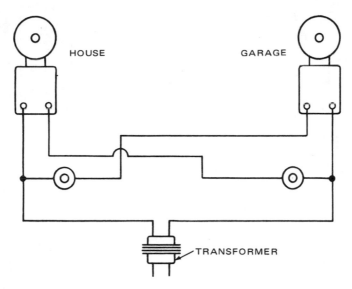

Fig. 12-22. Hookup for three-wire return signal system.

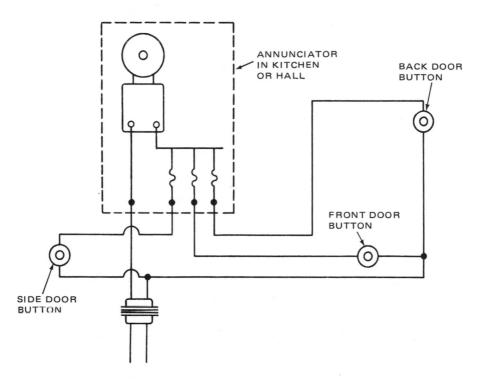

Fig. 12-23. Door bell system with three stations.

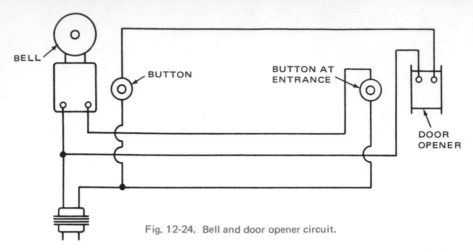

Fig. 12-24. Bell and door opener circuit.

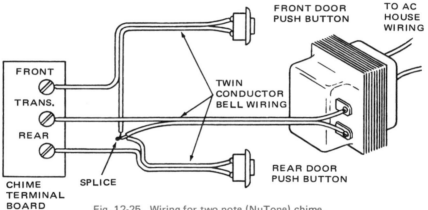

Fig. 12-25. Wiring for two note (NuTone) chime.

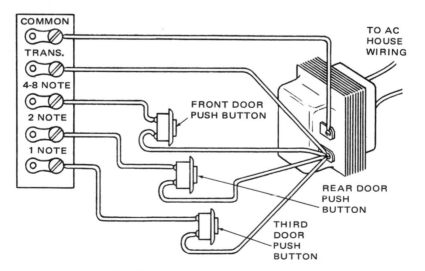

Fig. 12-26. Wiring for eight note chime.

House Wiring Circuits

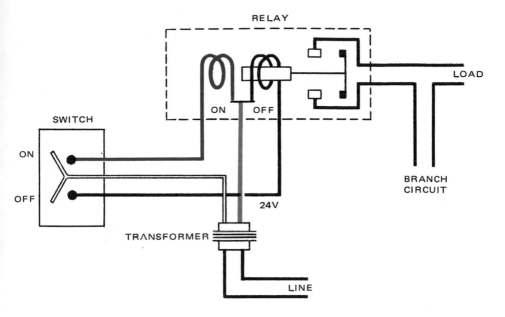

Fig. 12-27. Basic circuit for General Electric's remote-control, low-voltage switching system. Relays handle the switching of the current. The relays are controlled by switches operating at low voltage that permit the use of wiring similar to that used for door chimes. Many different components are available. In planning and installing remote control systems, instructions supplied by the manufacturer of the equipment being used should be followed.

House Wiring Simplified

Unit 13

MODERNIZING ELECTRICAL SYSTEMS

Installing electrical wiring in a home while it is being constructed is called NEW work. Providing wiring needed to install switches, outlets and fixtures in a home that is already built is called OLD work.

In both new work and old work the same basic wiring principles are involved.

Wiring in new work is largely a matter of mounting metal boxes on the framework and running conduit or cables to the boxes following the most direct route. When conduit is used, the wires are fished through the conduit and the necessary connections made after the walls are completed.

Old work, where the wires are to be concealed, involves getting cables and wires from one point to another with the least effort, and minimum damaging of structural members and finished walls.

Armored (BX) cable, plastic cable, and nonmetallic sheathed cable are used extensively for old concealed wiring, because they are flexible and may be easily pulled through openings between walls, ceilings and floors. The National Electric Code prohibits running flexible cord through holes in walls, ceilings, floors, etc. and using it as a substitute for cable.

Old work frequently requires more material than new work, as it is easier to run the cable through channels that are readily accessible, rather than tear up ceilings, floors and walls.

Solving problems that arise obviously requires some ingenuity on the part of the electrician handling the job.

CHECK CODES

In doing a job of modernizing a wiring system, check requirements of the local building Codes and Ordinances and see to it that all requirements are met. It would also be well to obtain a copy of the National Electrical Code and have it available for ready reference.

SAFETY: BE SURE THE ELECTRICITY IS DISCONNECTED BEFORE DISCONNECTING ANY WIRES OR MAKING WIRING HOOKUPS. DON'T TAKE CHANCES.

INSTALLING NEW OUTLETS

First decide where the new outlets are to be located, then figure out the best way to run cables to them. One end of the new cable must be run to an existing outlet box which contains a black wire (hot) and a ground wire (white) and is not already overloaded with circuits. See Figs. 13-1 and 13-2. Another possibility is to bring power from the entrance panel . . . start a new circuit.

Keep in mind the fact that all wire splices must be enclosed in boxes, and that each switch and convenience outlet must be housed in a box. Each fixture must be attached to a box.

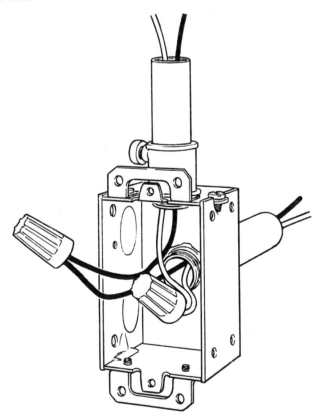

Fig. 13-1. When installing a new outlet the box tapped into must contain a black wire, which is continuously live, a white wire, and not be overloaded with circuits.

Box Dimensions, Inches Trade Size	Maximum Number of Conductors For Deep Boxes			
	No. 14	No. 12	No. 10	No. 8
3 1/4 x 1 1/2 Octagonal	5	4	4	3
3 1/2 x 1 1/2 Octagonal	5	5	4	3
4 x 1 1/2 Octagonal	8	7	6	5
4 x 2 1/8 Octagonal	11	10	9	7
4 x 1 1/2 Square	11	10	9	7
4 x 2 1/8 Square	15	14	12	10
4 11/16 x 1 1/2 Square	16	14	12	10
4 11/16 x 2 1/8 Square	23	20	18	15
3 x 2 x 1 1/2 Rectangular	3	3	3	2
3 x 2 x 2 Rectangular	5	4	4	3
3 x 2 x 2 1/4 Rectangular	5	5	4	3
3 x 2 x 2 1/2 Rectangular	6	5	5	4
3 x 2 x 2 3/4 Rectangular	7	6	5	4
3 x 2 x 3 1/2 Rectangular	9	8	7	6
4 x 2 1/8 x 1 1/2 Rectangular . . .	5	4	4	3
4 x 2 1/8 x 1 7/8 Rectangular . . .	6	6	5	4
4 x 2 1/8 x 2 1/8 Rectangular . . .	7	6	6	5

Box Dimensions, Inches Trade Size	Maximum Number of Conductors for Shallow Boxes		
	No. 14	No. 12	No. 10
3 1/4	4	4	3
4	6	6	4
1 1/4 x 4 Square	9	7	6
4 11/16	8	6	6

Fig. 13-2. Above. Maximum number of conductors which may be used in deep boxes. Below. Maximum number of conductors which may be used in shallow box (box less than 1 1/2 in. deep).

In old work rectangular boxes are generally used because they are relatively easy to install with minimum marring of the walls.

In old wiring where the system is grounded back to the entrance service panel (this is usually the case where conduit or armored cable have been used) outlet receptacles of the grounding type must be used. These have two parallel slots, a U-shaped opening for the prong on a 3-wire plug, and a green colored grounding terminal, Fig.

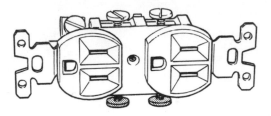

Fig. 13-3. Outlet receptacle with two parallel slots,
U-shaped opening, and grounding terminal.

13-3. A grounding wire must be run from the green terminal of the receptacle to the box. The wire may be fastened to the box using a sheet metal screw inserted in a hole provided for the purpose or with a metal clip, Fig. 13-4.

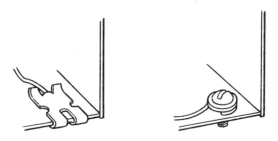

Fig. 13-4. Connecting grounding wire to metal box.

BACK TO BACK OUTLETS

Fig. 13-5 shows how a new outlet may be installed in the wall back of an existing outlet.

Decide location of box. Measure from door opening and baseboard. Drill a small hole (1/16 in.) through the wall and use a piece of wire as a probe to make sure there are no obstructions which will interfere with the installation.

Use a template for wall marking as shown in Figs. 13-6 and 13-7.

If the wall is lath and plaster or of other material which will not hold nails an outlet box of the type which has screw type clamps on the side may be used, Fig. 13-8. Or, you may use metal supports to hold a box of standard type, Fig. 13-9.

If the wall is lath and plaster, chip away plaster to determine lath locations. Then, cut one full lath and notch the top and bottom laths, Fig. 13-10.

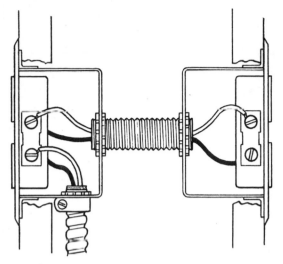

Fig. 13-5. Installing new outlet in wall back of old outlet.

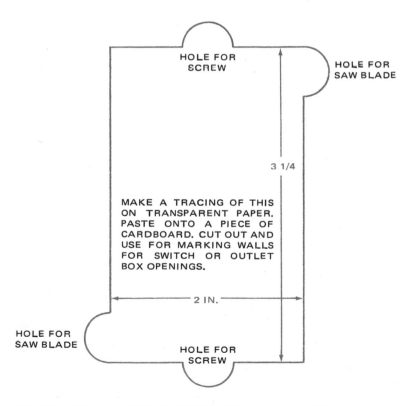

HOLE FOR
SCREW

HOLE FOR
SAW BLADE

3 1/4

MAKE A TRACING OF THIS
ON TRANSPARENT PAPER.
PASTE ONTO A PIECE OF
CARDBOARD. CUT OUT AND
USE FOR MARKING WALLS
FOR SWITCH OR OUTLET
BOX OPENINGS.

2 IN.

HOLE FOR
SAW BLADE

HOLE FOR
SCREW

Fig. 13-6. Full-size template for marking wall area opening to take outlet box.

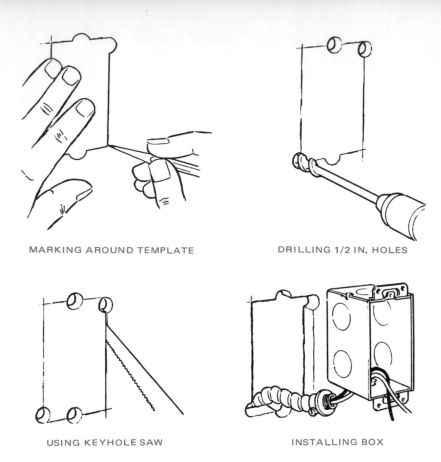

MARKING AROUND TEMPLATE

DRILLING 1/2 IN. HOLES

USING KEYHOLE SAW

INSTALLING BOX

Fig. 13-7. Preparing wall opening for outlet box.

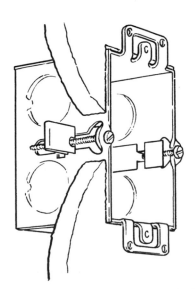

Fig. 13-8. Installing box with side brackets in dry wall. Side screws are tightened to bring the brackets against the wall and hold the box in place.

Modernizing Electrical Systems

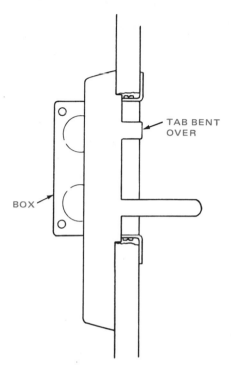

Fig. 13-9. Using metal supports to hold box in place.

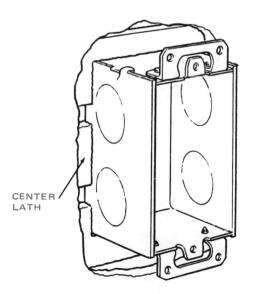

Fig. 13-10. Cut through center lath, and notch laths above and below.

INSTALLING NEW OUTLETS

Fig. 13-11 shows how to run a cable from a wall switch (provided neutral ground white wire is available) to new outlet installed above the baseboard. Remove the baseboard, chip plaster to make channel for BX cable as indicated. All bends must be made so the cable is not injured. The radius of the inner edge should not be less than five times the diameter of the cable. No splices in the cable are permitted between boxes.

If you have a job of installing an outlet on the first floor, and you can run the BX cable across the basement, proceed as indicated in Figs. 13-12 and 13-13. Use a long-shank bit to bore a hole at an angle.

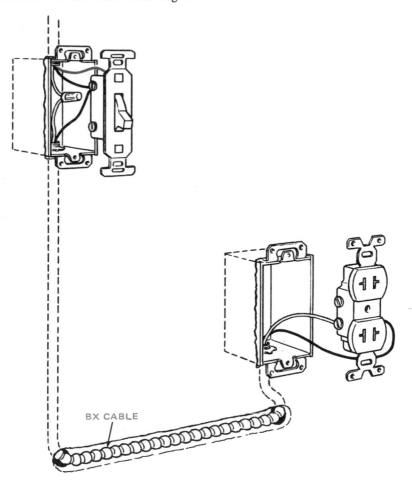

BX CABLE

Fig. 13-11. Running cable from wall switch to new outlet above baseboard.

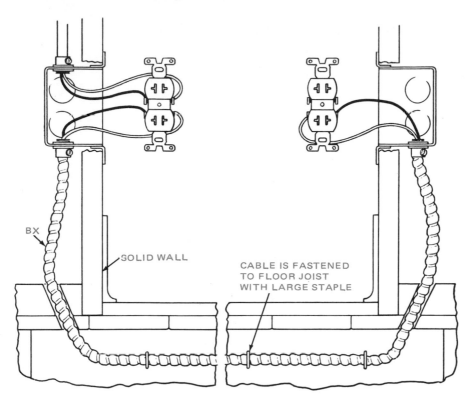

Fig. 13-12. Installing outlet by running cable through floor and across basement.

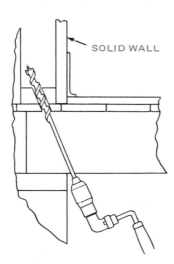

Fig. 13-13. Boring hole diagonally from basement through floor
so cable may be run between walls.

BASEMENT LIGHT TO FIRST FLOOR OUTLET

Fig. 13-14 illustrates the procedure to follow in running a cable from a basement light (with switch at light and ground wire in box) to a new outlet.

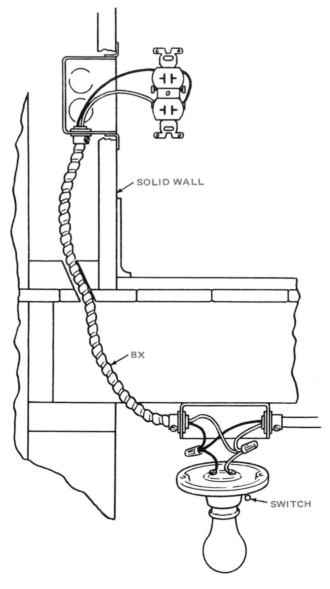

SOLID WALL

BX

SWITCH

Fig. 13-14. Running cable from basement light to new outlet on first floor.

RUNNING CABLE AROUND DOOR

To run a cable around a door and install an outlet on the other side of the door, refer to Fig. 13-15. Remove the door trim and a section of baseboard. Notch spacers to take BX cable. Run cable, then replace door trim and baseboard.

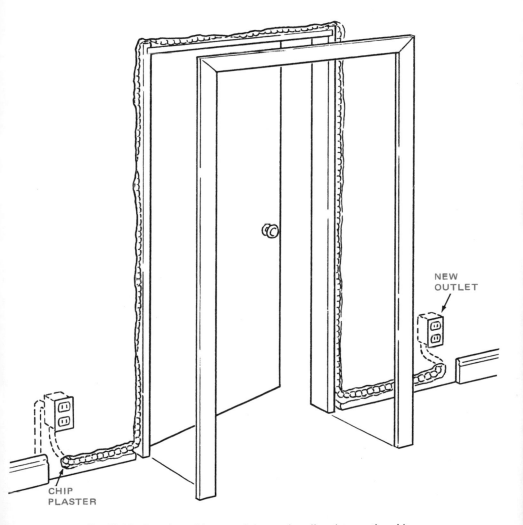

Fig. 13-15. Running cable around door to install outlet on other side.

INSTALLING WALL SWITCH

Installing a wall switch for a ceiling light that has been controlled previously by a pull switch at the fixture, is shown in Fig. 13-16. A two-wire cable (one black wire, one white) is run from the light to the box in which the new switch is installed. Be sure to mark both ends of the white wire with black paint so others will know both wires are hot. Connecting a black wire to a white wire is permitted by the Code if cable is used (not permitted in using conduit).

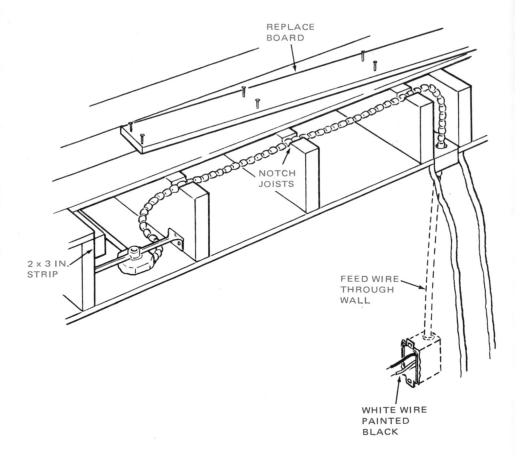

Fig. 13-16. Installing switch for ceiling light.

FISHING CABLE THROUGH WALL

Fishing a cable from the attic or room above the new installation is shown in Fig. 13-17. Remove baseboard and drill a hole diagonally downward as indicated in the top drawing. Push fish wires with hooks at ends through opening. Withdraw one wire until it hooks the other (lower drawing, Fig. 13-17), then withdraw second wire until hooks meet. Attach to wires in BX cable and pull cable through opening.

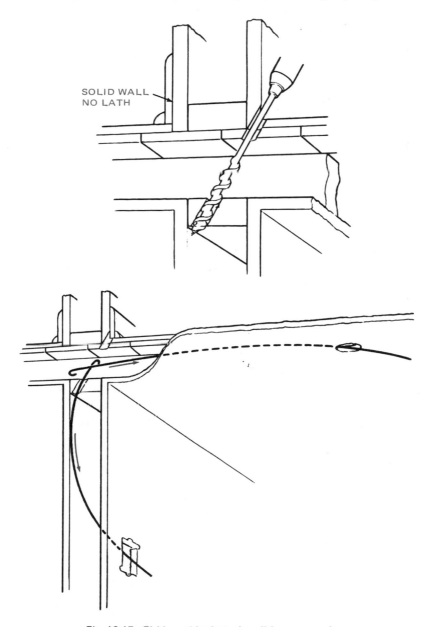

SOLID WALL
NO LATH

Fig. 13-17. Fishing cable through wall from room above.

LIFTING ATTIC FLOOR BOARD

The drawings, Figs. 13-18 and 13-19, show lifting attic floor boards, and notching the joists to take a cable. See where joist is nailed, then bore small holes (1/16 in.) to locate edge of joist. If flooring is of the tongue and groove type you may use putty knife with the edge sharpened and a hammer to cut through the tongue on both sides

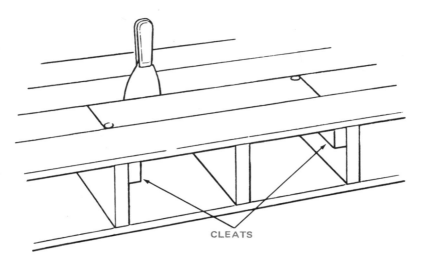

Fig. 13-18. Preparing section of attic flooring board for removal.

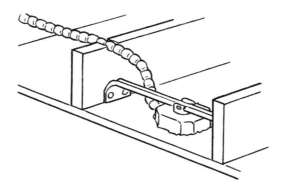

Fig. 13-19. Attic floor joists are notched to take cable.

of the portion to be removed. Bore holes at the corners large enough to take a small keyhole saw blade and saw through the board in two places. Notch the joists to take the cable, Fig. 13-19, and nail a cleat to the joists to support the floor board when it is replaced, Fig. 13-18.

Modernizing Electrical Systems

    Installing an octagon box is shown in Fig. 13-20. In old work where it is impractical to install an octagon shaped box of regular depth, a shallow round box and an old work hanger, may be used, Fig. 13-21. Round boxes must not be used where conduits or connectors requiring the use of locknuts or bushings are to be connected to the side of the box.

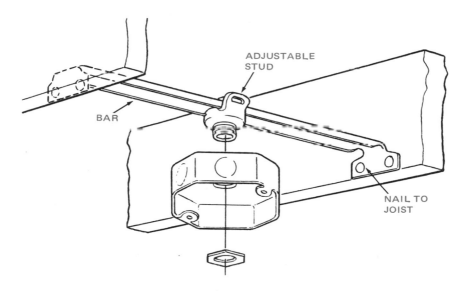

Fig. 13-20. Installing octagon box in ceiling.

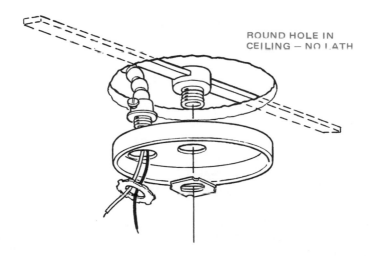

Fig. 13-21. Old work. Installing shallow, round box, using special old work hanger.

SURFACE WIRING

In moderizing house wiring the use of some surface wiring devices in inconspicuous areas, a real time saver, should be considered. See Fig. 13-22.

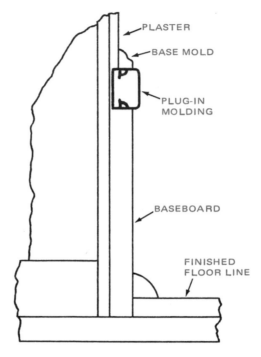

PLASTER

BASE MOLD

PLUG-IN MOLDING

BASEBOARD

FINISHED FLOOR LINE

Fig. 13-22. Using surface mounted plug-in molding.

OUTDOOR WIRING

Installing a yard light (typical installation) is shown in Fig. 13-23. The light is switch controlled from the house. The control may be manual, or an automatic time switch may be provided. The convenience outlets are always on. An easy way to provide the necessary wiring from the house to the light is to use dual-purpose plastic cable. Check local Codes before proceeding. It may be that the cable must be encased in conduit, or that lead-covered cable will be required. Keep in mind the fact that the principles involved in outdoor wiring and indoor wiring are basically the same.

LOW VOLTAGE EQUIPMENT: Typical low voltage (6-12V) lighting equipment designed for outdoor use is shown in Fig. 13-24. This is used in connection with regular 120 voltage equipment. Low voltage lighting is ideal for lighting small gardens and for providing specific accents in larger gardens.

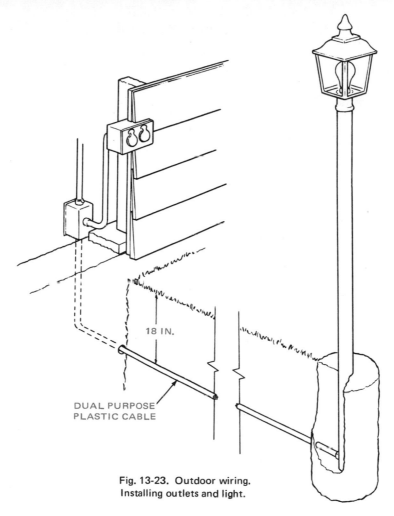

18 IN.

DUAL PURPOSE
PLASTIC CABLE

Fig. 13-23. Outdoor wiring.
Installing outlets and light.

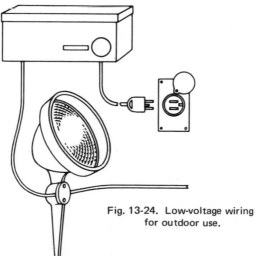

Fig. 13-24. Low-voltage wiring
for outdoor use.

WINTER ROOF DRAINAGE

Using electric heating cable to prevent downspout and gutter freezeup, is shown in Fig. 13-25. Lay cable in roof gutters, drop end in downspout, then fasten cable to roof with special clips provided. Electricity required may be provided by installing an outdoor type outlet box, or by using a heavy-duty extension cord run from a nearby outdoor outlet.

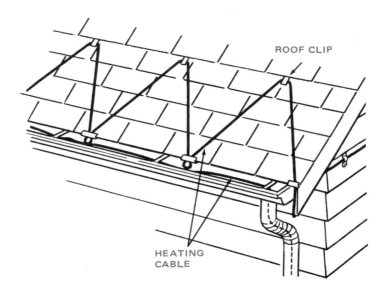

Fig. 13-25. Using heating cable to prevent gutter and downspout freezeup.

PROGRESS CHECK — UNIT 13

1. Different basic wiring principles are involved in wiring new work and in wiring old work. True or False?
2. Three kinds of cable used extensively for old concealed wiring are:
   a. _____ .
   b. _____ .
   c. _____ .
3. When installing new outlets, one end of the new cable must run to an existing outlet box which contains a_____wire and a_____wire and which is not _____ .
4. What type of outlet receptacles must be used in old wiring where the system is

grounded back to the entrance service panel?

5. For marking wall openings to take boxes, use a _____ of the same size.

6. When bending BX cable, the radius of the inner edge should not be less than _____ times the diameter of the cable.

7. In modernizing, when a two-wire cable is run from a light to a box in which a new switch is to be installed, be sure to mark both ends of the_____wire so others will know it is_____.

House Wiring Simplified

Unit 14

USEFUL INFORMATION

EXTENSION CORDS

Extension cords are used to reach outlets that are not close enough to plug appliances into directly. These cords are made up of fine strands of copper, so they will be flexible. Each cord has a maximum allowable current-carrying capacity, Fig. 14-1. Appliances and power tools which have grounding plugs should be used only with 3-wire grounding type extension cords and plugged into 3-hole grounding outlets.

Damaged or frayed cords provide serious fire hazards and should be replaced. Exception: A cord that is frayed only at one end may sometimes be repaired. Cut cord to remove damaged portion, then attach wire ends to new plug.

ABILITY OF CORD TO CARRY CURRENT (2 OR 3-WIRE CORD)

Wire Size	Normal Load	Capacity Load
No. 18	5.0 Amp. (600W)	7 Amp. (840W)
No. 16	8.3 Amp. (1000W)	10 Amp. (1200W)
No. 14	12.5 Amp. (1500W)	15 Amp. (1800W)
No. 12	16.6 Amp. (1900W)	20 Amp. (2400W)

TYPES AND USE OF EXTENSION CORDS

	Wire Size	Use
Ordinary Lamp Cord	No. 16 or 18	In residences for lamps or small appliances.
Heavy-duty — with thicker covering	No. 10, 12, 14 or 16	In shops, and outdoors for larger motors, lawn mowers, outdoor lighting, etc.

SELECTING LENGTH OF CORD

Light Load (to 7 Amps.)	Medium Load (7-10 Amps.)	Heavy Load (10-15 Amps.)
To 25 Ft.—Use No. 18	To 25 Ft.—Use No. 16	To 25 Ft.—Use No. 14
To 50 Ft.—Use No. 16	To 50 Ft.—Use No. 14	To 50 Ft.—Use No. 12
To 100 Ft.—Use No. 14	To 100 Ft.—Use No. 12	To 100 Ft.—Use No. 10

Fig. 14-1. Selecting proper type of extension cord for job at hand.

READING ELECTRIC METERS

Electric meters register in kilowatt-hour units. A killowatt-hour is 1,000 watts in operation for one hour.

Fig. 14-2 shows how the four dials look on a typical meter. Arrows indicate the direction the hands rotate.

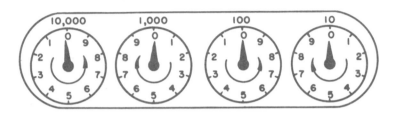

Fig. 14-2. Dials of a typical electric meter. The arrows indicate the direction hands on the dials rotate.

To take a reading you must read all four dials of the meter. The figures on the dial at the extreme right measure individual kilowatt-hours. Each figure on the second dial from the right shows 10 kilowatt-hours. Each figure on the third dial from the right represents 100 kilowatt-hours. Each figure on the dial at the left represents 1,000 kilowatt-hours.

To fix these values in your mind, remember that the dial at the extreme right is capable of registering 10 kilowatt-hours. The second dial from the right is capable of registering 100 kilowatt-hours. The third dial from the right is capable of registering 1,000 kilowatt-hours. The dial at the left will register 10,000 kilowatt-hours.

In each case, the last figure passed by the hand, and not the nearest, is used in the reading. When the hand seems to be right on the number, the dial at the right should be consulted to see whether or not the number has been passed.

In reading a meter, always start with the dial on the right and remember that the pointer on the right dial must make one complete revolution before the indicator on the next dial moves one number.

The kilowatt-hours used in a month are determined by subtracting the reading at the start of the month from the end-of-the-month reading. Before you can figure the cost of the electrical energy used, you must obtain information on rates from the utility company. You will find that the cost per kilowatt-hour goes down as you use more electricity.

Useful Information

BULB
Soft glass is generally used. Hard glass is used for some lamps to withstand higher bulb temperatures and for added protection against bulb breakage due to moisture. Bulbs are made in various shapes and finishes.

GAS
Usually a mixture of nitrogen and argon is used in most lamps 40 watts and over to retard evaporation of the filament.

FILAMENT
The filament material generally used is tungsten. The filament may be a straight wire, a coil or a coiled-coil.

SUPPORT WIRES
Molybdenum wires support the filament.

LEAD-IN WIRES
Made of copper from base to stem press and nickel-plated copper or nickel from stem press to filament; carry the current to and from the filament.

BUTTON
Glass is heated during manufacture and support and tie wires placed in it.

BUTTON ROD
Glass rod supports button.

TIE WIRES
Molybdenum wires support lead-in wires.

HEAT DEFLECTOR
Used in higher wattage general service lamps and other types when needed, to reduce circulation of hot gases into neck of bulb.

STEM PRESS
The lead-in wires in the glass have an air-tight seal here and are made of a combination of a nickel-iron alloy core and a copper sleeve (Dumet wire) to assure about the same coefficient of expansion as the glass.

FUSE
Protects the lamp and circuit by blowing if the filament arcs.

EXHAUST TUBE
Air is exhausted through this tube during manufacture and inert gases introduced into the bulb. The tube, which originally projects beyond the bulb, is then sealed off short enough to be capped by the base.

BASE
Typical screw base is shown. One lead-in wire is soldered to the center contact and the other soldered or welded to the upper rim of the base shell. Made of brass or aluminum.

Fig. 14-3. Typical incandescent lamp bulb. This type produces a high lighting level over a relatively long period of time. Longer lasting lamps can be produced but the light output is lower. Additional light is produced at the expense of lamp life. Modern incandescent lamps strike a balance between light intensity and lamp life. Bulb blackening is the result of depositing of tungsten particles on inner surface of the bulb.   (Sylvania)

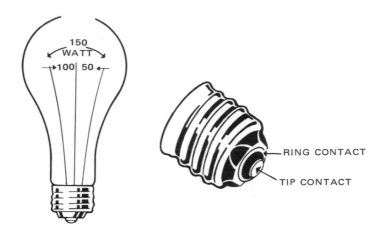

Fig. 14-4. Typical three-way bulbs have two filaments. Each filament may be operated separately, or in combination with the other. First, the low-wattage filament is switched on, next the high-wattage filament, then the two are switched on together. A special three-way socket with two contacts in the base, and a three-way switch are required.   (General Electric)

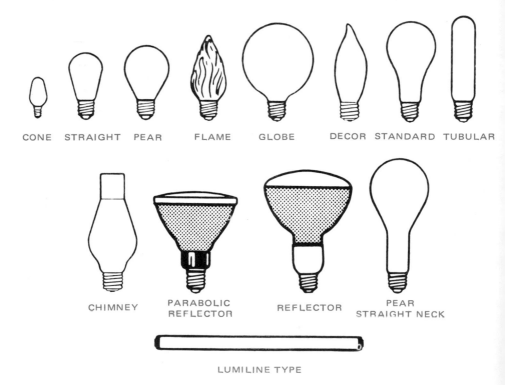

Fig. 14-5. Incandescent lamp bulb shapes.

Useful Information

Mini-Can
Screw

Candelabra
Cand.

Intermediate
Inter.

Double Contact
Bayonet
Candelabra
D. C. Bay.

Disc
(Lumiline)

Medium
Med.

3 Kon-Tact
Medium
3 C. Med.

Medium
Prefocus
Med. Pf.

Medium Side Prong

Mogul
Mog.

Three Contact
Mogul
3 C. Mog.

Mogul
Prefocus
Mog. Pf.

Medium
Skirted
Med. Skt.
(Mechanical)

Medium
Skirted
Med. Skt.
(Cement)

Fig. 14-6. Bases for incandescent lamps.

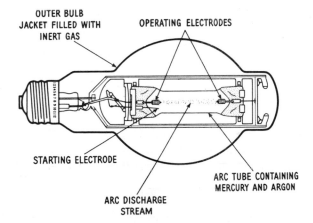

Fig. 14-7. Mercury lamp which produces light by passing an alterating current through mercury vapor in arc tube.

131

Fig. 14-8. Test lamp you can make. Use two pieces of insulated wire, No. 14, about 15 in. long. Connect one end of wires to nonmetallic socket. Remove insulation from other ends of wire and wrap with tape, as shown. Use lamp of low wattage.

FLUORESCENT LAMPS

The fluorescent lamp, Fig. 14-9, is a long narrow or circular glass cylinder, coated on the interior with any of several types of phosphor (chemical coating which will radiate light). Air in the tube is replaced with mercury vapor and argon, an inert or chemically inactive gas. At each end of the lamp is an electrode, made of an oxide-coated tungsten filament. When heated by an electric current, the filament releases a cloud of electrons around each electrode.

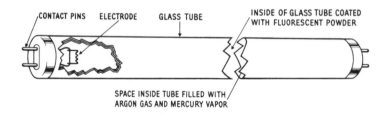

Fig. 14-9. Fluoescent lamp.

A high voltage electrical surge then establishes an electron arc between the electrodes with each alternation of the current. The electrons collide with mercury vapor and argon gas atoms filling the tube to produce invisible ultraviolet rays.

The rays excite the fluorescent phosphor coating the inside of the tube to become visible light.

Unlike incandescent lamps, fluorescent lamps cannot control their current consumption. Unless the current is controlled, the lamp burns itself out immediately. To avoid this, ballasts, (devices which limit current to the proper operating value) are wired into the fluorescent circuit.

Small fluorescent lamps require choke ballast to limit current input to prevent burnout. Large lamps require both a choke coil and a transformer. The transformer steps up the voltage and the coil limits the current.

Useful Information

SIMPLE SPECIFICATION FORM

The following material is intended as a guide to preparing wiring specifications for dwellings.

All outlets, the locations of wall switches, and the outlet or outlets controlled by each switch should be shown clearly on the floor plans which must be considered as an essential part of the plans, or contract.

SPECIFICATIONS FOR ELECTRIC WIRING IN THE DWELLING
TO BE ERECTED AT_____ FOR_____

1. GENERAL

The installation of electric wiring and equipment shall conform with local regulations, the National Electrical Code, and the requirements of the local electric service company. All materials shall be new and shall be listed by Underwriters' Laboratories, Incorporated, as conforming to its standards, in every case where such a standard has been established for the particular type of material in question.

2. GUARANTEE

The contractor shall leave his work in proper order and, without additional charge, replace any work or material which develops defects, except from ordinary wear and tear, within one year from the date of the final certificate of approval.

3. WIRING METHODS

Interior wiring shall be _____. No exposed wiring shall be installed except in unfinished portions of basement, utility room, garage, attic, and other spaces that may be unfinished.

4. SERVICE ENTRANCE conductors shall be three No._____wires.

_____
(Fill in wiring method)
5. SERVICE-EQUIPMENT shall consist of_____
_____

6. TWENTY-AMPERE BRANCH CIRCUITS

At least_____20-ampere branch circuits shall be installed to supply all lighting outlets and all convenience outlets except those which are supplied by appliance branch circuits. The total number of outlets shall as nearly as possible be divided equally between these circuits. In each living room, library, sun room, bedroom, and each other principal room, the outlets shall be divided between two or more branch circuits.

7. APPLIANCE BRANCH CIRCUITS

Two appliance branch circuits shall be installed to supply all convenience outlets in the dining room, breakfast room, kitchen and pantry (except clock outlet), and laundry. These circuits shall be so installed that convenience outlets served by both circuits will be available in both the kitchen and the laundry.

House Wiring Simplified

8. BRANCH CIRCUIT EQUIPMENT shall be _____

_____
(Fill in type of equipment)

9. OUTLETS AND SWITCHES

Lighting outlets, convenience outlets complete with receptacles, and switches shall be installed as shown on the plans.

Unless otherwise shown on plans, the height of outlets above floor shall be approximately:
Switches .48 inches
Convenience Outlets . . . . .18 inches

10.SPECIAL PURPOSE OUTLETS AND CIRCUITS shall be installed as shown on the plans. The circuits shall be:

Circuits for	No. of Wires	Size
_____	_____	_____
_____	_____	_____
_____	_____	_____
_____	_____	_____

Useful Information

AN ORDINANCE PROVIDING SAFE AND PRACTICAL STANDARDS
AND SPECIFICATIONS FOR INSTALLATION, ALTERATION
AND USE OF ELECTRICAL EQUIPMENT.

PERMITS REQUIRED: No electrical wiring for light, heat or power shall be started or installed in any building or structure or on any premises within the Village of South Holland, nor shall any alteration or extension of any existing electric wiring system be made, nor shall any electrical equipment or fixtures hereinafter described be installed, changed, moved, relocated or altered until a permit therefore has first been obtained as provided herein, and no such wiring, installation or equipment shall be used or operated until finally inspected and approved by the Chief Electrical Inspector for the Village of South Holland, or his duly authorized deputy.

All such wiring, equipment and the installation thereof shall conform to all the specifications, rules and regulations contained in this ordinance, and no permit shall be issued unless plans showing conformity with all provisions of this ordinance have been submitted and approved as hereinafter provided.

UNDERWRITERS' LABORATORIES INC. APPROVAL: All materials, devices, conductors, enclosures, conduits, major household appliances, lighting fixtures or any other electrical current consuming device to be permitted, shall bear the seal of approval of the listings of the Underwriters' Laboratories, Inc.

SERVICE ENTRANCE CAPACITY AND EQUIPMENT REQUIREMENT: The minimum service capacity approved for any dwelling occupancy shall be a 3-wire 120/240 volt, solid neutral, 100 ampere service. Service entrance conductors shall not have a lesser capacity than 3 No. 3 type R.H., or 2 No. 3 and 1 No. 4 type R.H. copper conductors.

Service entrance conductors shall be installed in rigid metallic conduit.

Residential occupancy service entrance capacity shall be calculated on the basis of 100 watts for each convenience, ceiling, or other outlet, plus the total connected load of all appliances, motors, or other electrical equipment.

SERVICE ENTRANCE PANELS: Service entrance panels shall have not less than 12 circuit positions. Panels approved may be circuit breaker type with main disconnect, or fused type panels with main disconnect.

MAIN AND BRANCH CIRCUIT DISCONNECTS: Any disconnecting devices for motors, heating equipment, air conditioning, hot water heaters, or any other major appliances shall be DEAD FRONT, enclosed Circuit Breaker, Fused pullout, or door operated type.

EDISON-BASE TYPE PLUG FUSES: All Fuse Plugs shall be (S) type in accord with Section 240-20 National Electrical Code as from time to time amended.

CONDUCTORS — GENERAL: The use of aluminum conductors is prohibited in residential occupancies at any point beyond the line side terminals of the service entrance panel.

DIRECT BURIAL: The use of any type wire other than that approved for direct burial (type UFE) is prohibited for underground services, and for any other purpose.

RACEWAYS OR CONDUITS. NEW WORK: For all new work, wiring shall be installed in rigid conduit, electrical metallic tubing, or surface metal raceways, except that it shall be permissible to use a section of flexible Greenfield metal conduit, where flexible connections are necessary, not exceeding 3 feet in length. The use of armored flexible cable (BX) is prohibited on new work.

House Wiring Simplified

INDIVIDUAL BRANCH CIRCUITS: Individual branch circuits shall be provided in dwelling occupancies as required for general purpose circuits calculated on the basis of five (5) watts per square foot, and in addition, individual branch circuits for:

Heating plants	Well pumps
Electric ranges	Sump or sewerage ejection pumps
Electric ovens	Heat pumps
Room air conditioning	Space heating, cable, radiant, convection
units larger than one ton	or unit type heaters
Fixed air conditioners	Garbage grinders
Electric space heaters	Dishwashers
1000 Watts and larger	Any other major electric appliance
Electric water heaters	not listed above

RECEPTACLE OUTLETS REQUIRED: Dwelling Type Occupancies. (N.E. Code 210-22). In every kitchen, dining room, breakfast room, living room, parlor, library, den, sun room, recreation room, and bedroom, receptacle outlets shall be installed so that no point along the floor line in any usable wall space is more than six feet, measured horizontally, from an outlet in that space including any usable wall space two feet wide or greater and the wall space occupied by sliding panels in exterior walls. The receptacle outlets shall, insofar as practicable, be spaced equal distances apart. Garages shall be wired and have a minimum of one ceiling outlet and one convenient wall receptacle. All receptacles shall be of the Grounding type.

CONVENIENCE AND CEILING OUTLETS, GENERAL PURPOSE CIRCUITS: The total number of these outlets on a No. 14 circuit shall not exceed (10) outlets. On a No. 12 circuit the total number of these outlets shall not exceed (16) outlets. A minimum of 2 — No. 12 circuits are required in the kitchen and 1 — No. 12 circuit in utility for wall receptacles.

PERMANENT LIGHTING FIXTURES: 1—Permanent lighting fixtures, wall switch controlled, shall be installed in kitchens, bath rooms, powder rooms, halls and other applicable locations. 2—Where there is more than one entrance to the above indicated rooms and/or hallways, permanent lighting fixtures shall be multiple switch controlled at both entrances. 3—A permanent lighting fixture shall be installed to illuminate properly all basement stairs, and such fixture shall be switch controlled from the head of the stairs. Where there is more than one entrance to the basement, then the permanent lighting fixture shall be multiple switch controlled from both entrances. 4—A minimum of one W/P outlet shall be installed on the outside of all dwelling type occupancies.

OUTLET BOXES AND COVERS IN WOOD PANELING, SIDING PLYWOOD, OR OTHER COMBUSTIBLE MATERIALS. OUTLET BOXES IN DRY WALL ACOUSTICAL TILE, OR SIMILAR MATERIALS: 1—IN WALL OR CEILING. In walls or ceilings of concrete, tile or other non-combustible material, boxes and fittings shall be so installed that the front edge of the box or fitting will not set back of the finished surface more than 1/4 inch. In walls and ceilings constructed of wood or other combustible material, outlet boxes and fittings shall be flush with the finished surface or project therefrom. **2—REPAIRING PLASTER.** On walls or ceilings of concrete, tile or other non-combustible material, a plaster surface which is broken or incomplete shall be repaired so that there will be no gaps or open spaces at the edge of the box or fitting. **3—NO ELECTRICAL LIGHTING FIXTURES** shall be hung until all such wall material has been repaired.

PLANS AND SPECIFICATIONS: Where construction or alteration work is to be done in single family dwelling occupancies, the plans shall include wiring layout showing:

1. Number and location of outlets.
2. Size of wire and conduit to be used.
3. Number of circuits and their ampere rating.

Useful Information

4. Location of service entrance equipment.
5. Location of major appliance.
6. Size of existing old service.
7. Electrically heated homes shall have location and wattage of all heating equipment shown, and circuits listed for same.

Where construction or alteration work is to be done in other than single family dwelling occupancies, the applicant shall submit with the application for permit, a complete wiring plan for approval. This plan shall indicate size of service conductors, the type and location of all outlets, the sizes of all meters and power consuming equipment; i.e., all current consuming devices which shall make up the total connected load. Conduit runs, size of conductors for branch circuits and feeders to be installed. The location of all panel boards and cabinets and the number and ampere rating of the circuits shall be indicated.

INSPECTION: It shall be the responsibility of the person, firm, or corporation doing the installation work, to notify the Village Electrical Department which shall make a roughing-in, trlm, and final inspection to assure compliance with the provisions of this ordinance.

ADOPTION OF NATIONAL ELECTRICAL CODE: The National Electrical Code, being the Standard of the National Board of Fire Underwriters for Electric Wiring and Apparatus as Recommended by the National Fire Protection Association, approved by American Standards Association, copies of which have been placed on file for more than thirty days last past with the Village Clerk of the Village of South Holland, is hereby incorporated herein and adopted as to all matters therein covered which are not specifically covered by this ordinance.

CONFLICTING ORDINANCES: All ordinances or parts of ordinances in conflict herewith, are hereby repealed to the extent of such conflict.

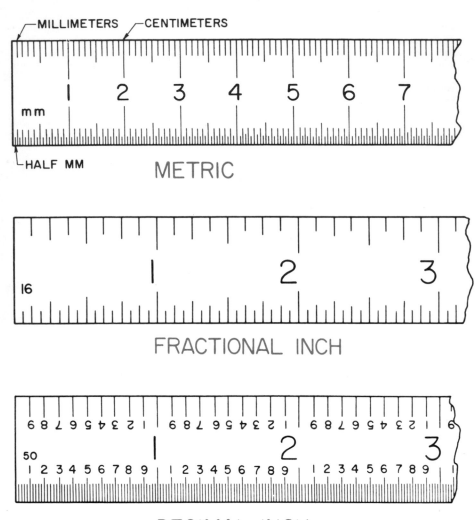

Fig. 14-10. Ninety percent of the countries of the world use the metric system. The National Bureau of Standards has recommended to Congress that the U. S. change over to the metric system through a carefully planned program over a period of years. This drawing compares a metric (millimeter) rule with the conventional fractional and decimal inch rules. For conversion; 1 yard = 0.914 meters, 1 foot = 30.480 centimeters, and 1 inch = 2.540 centimeters. Also, for conversion; 1 meter = 3.281 feet, and 1 centimeter = 0.3937 inch.

Table 210-25
Branch-Circuit Requirements

(Type FEP, FEPB, RUW, SA, T, TW, RH, RUH, RHW, RHH, THHN, THW, THWN, and XHHW conductors in raceway or cable.)

CIRCUIT RATING	15 Amp.	20 Amp.	30 Amp.	40 Amp.	50 Amp.
CONDUCTORS: (Min. Size)					
Circuit Wires*	14	12	10	8	6
Taps	14	14	14	12	12
Fixture Wires and Cords	Refer to Section 240-5(a), Exception No. 2				
OVERCURRENT PROTECTION	15 Amp.	20 Amp.	30 Amp.	40 Amp.	50 Amp.
OUTLET DEVICES:					
Lampholders Permitted	Any Type	Any Type	Heavy Duty	Heavy Duty	Heavy Duty
Receptacle Rating**	15 Max. Amp.	15 or 20 Amp.	30 Amp.	40 or 50 Amp.	50 Amp.
MAXIMUM LOAD	15 Amp.	20 Amp.	30 Amp.	40 Amp.	50 Amp.
PERMISSIBLE LOAD	Refer to Section 210-24(a)	Refer to Section 210-24(a)	Refer to Section 210-24(b)	Refer to Section 210-24(c)	Refer to Section 210-24(d)

* These ampacities are for copper conductors where derating is not required. See Tables 310-12 through 310-15.

** For receptacle rating of cord-connected electric-discharge lighting fixtures see Section 410-14.

ARTICLE 220 — BRANCH-CIRCUIT AND FEEDER CALCULATIONS

220-1. Scope. This Article provides the basis for calculating the expected branch-circuit and feeder loads and for determining the number of branch circuits required.

220-2. Calculation of Branch-Circuit Loads. The load for branch circuits shall be computed in accordance with the provisions of this Section.

The continuous load supplied by a branch circuit shall not exceed 80 percent of the branch-circuit rating.

Exception No. 1: Where the assembly, including the overcurrent devices protecting the branch circuit and feeder is approved for operation at 100 percent of their rating, the continuous load supplied by the branch circuit may equal the ampacity of the branch-circuit conductors.

Exception No. 2: Where branch circuits are derated in accordance with Note 8 to Tables 310-12 through 310-15, branch-circuit loads shall not exceed the derated ampacity of the conductors.

(a) General Lighting Load.

(1) In Listed Occupancies. In the occupancies listed in Table 220-2(a), a load of not less than the unit load specified shall be included for each square foot of floor area.

House Wiring Simplified

In determining the load on the "watts per square foot" basis, the floor area shall be computed from the outside dimensions of the building, apartment or area involved, and the number of floors; not including open porches, garages in connection with dwelling occupancies, nor unfinished spaces and unused spaces in dwellings unless adaptable for future use.

The unit values herein are based on minimum load conditions and 100 percent power factor, and may not provide sufficient capacity for the installation contemplated.

(*continued on page 70-32*)

Table 220-2(a). General Lighting Loads by Occupancies

Type of Occupancy	Unit Load per Sq. Ft. (Watts)
Armories and Auditoriums	1
Banks	5
Barber Shops and Beauty Parlors	3
Churches	1
Clubs	2
Court Rooms	2
*Dwellings (Other Than Hotels)	3
Garages—Commercial (storage)	½
Hospitals	2
*Hotels and Motels, including apartment houses without provisions for cooking by tenants	2
Industrial Commercial (Loft) Buildings	2
Lodge Rooms	1½
Office Buildings	5
Restaurants	2
Schools	3
Stores	3
Warehouses Storage	¼
In any of the above occupancies except single-family dwellings and individual apartments of multi-family dwellings:	
Assembly Halls and Auditoriums	1
Halls, Corridors, Closets	½
Storage Spaces	¼

*All receptacle outlets of 15-ampere or less rating in single-family and multi-family dwellings and in guest rooms of hotels and motels [except those connected to the receptacle circuits specified in Section 220-3(b)] may be considered as outlets for general illumination, and no additional load need be included for such outlets.

140

(220-2(a-1) continued)

In view of the trend toward higher intensity lighting systems and increased loads due to more general use of fixed and portable appliances, each installation should be considered as to the load likely to be imposed and the capacity increased to insure safe operation.

Where electric-discharge lighting systems are to be installed, high-power-factor type should be used or the conductor capacity may need to be increased.

(2) In Other Occupancies. In other occupancies, a load of not less than the unit load specified in Section 220-2(b) shall be included for each outlet.

(b) Other Loads. For lighting other than general illumination and for appliances other than motors, a load of not less than the unit load specified below shall be included for each outlet. The loads indicated below are based on nominal branch-circuit voltages.

*Outlets supplying specific appliances and other loads ....
.... Ampere rating of appliance

Outlets supplying heavy-duty lampholders .... 600 volt-amperes

‡Other outlets .... 180 volt-amperes

    *For motors, see Sections 430-22 and 430-24.

‡This provision shall not be applicable to receptacle outlets connected to the circuit specified in Section 220-3(b) nor to receptacle outlets provided for the connection of stationary equipment as provided for in Section 400-3.

(c) Exceptions. The minimum load for outlets specified in Section 220-2(b) shall be modified as follows:

Exception No. 1: Ranges. For household electric ranges, the branch-circuit load may be computed in accordance with Table 220-5.

Exception No. 2: Show-Window Lighting. For show-window lighting a load of not less than 200 watts for each linear foot of show window, measured horizontally along its base, may be allowed in lieu of the specified load per outlet.

Exception No. 3: Multioutlet Assemblies. Where fixed multioutlet assemblies are employed, each 5 feet or fraction thereof of each separate and continuous length shall be considered as one outlet of not less than 1½ ampere capacity; except in locations where a number of appliances are likely to be used simultaneously, when each one foot or fraction thereof shall be considered as an outlet of not less than 1½ amperes. The requirements of this Section are not applicable to dwellings or the guest rooms of hotels.

Exception No. 4: Telephone Exchanges. Shall be waived for manual switchboards and switching frames in telephone exchanges.

The provisions of Section 220-2(b) shall apply to all other receptacle outlets.

(d) Existing Installations. Additions to existing installations shall conform to the following:

(1) Dwelling Occupancies. New circuits or extensions to existing circuits may be determined in accordance with Sections 220-2(a) or (b); except that portions of existing structures not previously wired, or additions to the building structure, either of which exceeds 500 square feet in area, shall be determined in accordance with Section 220-2(a).

(2) Other Than Dwelling Occupancies. When adding new circuits or extensions to existing circuits in other than dwelling occupancies, the provisions of Section 220-2(a) or (b) shall apply.

220-3. Branch Circuits Required. Branch circuits shall be installed as follows:

(a) Lighting and Appliance Circuits. For lighting, and for appliances, including motor-operated appliances, not specifically provided for in Section 220-3(b), branch circuits shall be provided for a computed load not less than that determined by Section 220-2.

The number of circuits shall be not less than that determined from the total computed load and the capacity of circuits to be used. In every case the number shall be sufficient for the actual load to be served, and the branch circuit loads shall not exceed the maximum loads specified in Section 210-23.

Where the load is computed on a "watts per square foot" basis, the total load, in so far as practical, shall be evenly proportioned among the branch circuits according to their capacity.

When lighting units to be installed operate at other than 100 percent power factor, see Section 210-23(b) for maximum ampere load permitted on branch circuits.

For general illumination in dwelling occupancies, it is recommended that not less than one branch circuit be installed for each 500 square feet of floor area in addition to the receptacle circuits called for in Section 220-3(b).

See Examples No. 1, 1a, 1b, 1c, and 4, Chapter 9.

(b) Small Appliance Branch Circuits, Dwelling Occupancies. For the small appliance load, including refrigeration equipment, in kitchen, pantry, family room, dining room, and breakfast room of dwelling occupancies, two or more 20-ampere appliance branch circuits in addition to the branch circuits specified in Section 220-3(a) shall be provided for all receptacle outlets in these rooms, and such circuits shall have no other outlets.

Receptacle outlets supplied by at least two appliance receptacle branch circuits shall be installed in the kitchen.

At least one 20-ampere branch circuit shall be provided for laundry receptacle(s) required in Section 210-22(b).

Receptacle outlets installed solely for the support of and the power supply for electric clocks may be installed on lighting branch circuits.

A 3-wire 115/230-volt branch circuit is the equivalent of two 115-volt receptacle branch circuits.

(c) Other Circuits. For specific loads not otherwise provided for in Section 220-3(a) or (b), branch circuits shall be as required by other sections of the Code.

220-4. Calculation of Feeder Loads. The computed load of a feeder shall be not less than the sum of all branch-circuit loads supplied by the feeder, as determined by Section 220-2, subject to the following provisions:

(Exception on page 70-34)

Useful Information

(220-4 continued)

Exception: When the calculated load for multi-family dwellings under this Section without electric cooking exceeds that calculated under Section 220-9 for the identical load plus electric cooking (based on 8 kW per unit), the lesser of the two loads may be used.

(a) Continuous and Noncontinuous Loads. When a feeder supplies continuous loads or any combination of continuous and noncontinuous load the rating of the overcurrent devices shall not be less than the noncontinuous load plus 125 percent of the continuous load.

Exception: When the assembly including the overcurrent devices protecting the feeder(s) are approved for operation at 100 percent of their rating, the ampacity of the feeder may equal the sum of the continuous load plus the noncontinuous load.

(b) General Lighting. The demand factors listed in Table 220-4(b) may be applied to that portion of the total branch-circuit load computed for general illumination. These demand factors shall not be applied in determining the number of branch circuits for general illumination supplied by the feeders.

See Sections 220-4(h) and (i).

The demand factors herein are based on minimum load conditions and 100 percent power factor, and in specific instances may not provide sufficient capacity for the installation contemplated. In view of the trend toward higher intensity lighting systems and increased loads due to more general use of fixed and portable appliances, each installation should be considered as to the load likely to be imposed and the capacity increased to insure safe operation. Where electric-discharge lighting systems are to be installed, high-power-factor type should be used or the conductor capacity may need to be increased.

**Table 220-5. Demand Loads for Household Electric Ranges,
Wall-Mounted Ovens, Counter-Mounted Cooking Units and
Other Household Cooking Appliances over 1¾ kW Rating**
**Column A to be used in all cases except as otherwise permitted
in Note 4 below.**

NUMBER OF APPLIANCES	Maximum Demand (See Notes)	Demand Factors (See Note 4)	
	COLUMN A (Not over 12 kW Rating)	COLUMN B (Less than 3½ kW Rating)	COLUMN C (3½ kw to 8¾ kW Rating)
1	8 kW	80%	80%
2	11 kW	75%	65%
3	14 kW	70%	55%
4	17 kW	66%	50%
5	20 kW	62%	45%
6	21 kW	59%	43%
7	22 kW	56%	40%
8	23 kW	53%	36%
9	24 kW	51%	35%
10	25 kW	49%	34%
11	26 kW	47%	32%
12	27 kW	45%	32%
13	28 kW	43%	32%
14	29 kW	41%	32%
15	30 kW	40%	32%
16	31 kW	39%	28%
17	32 kW	38%	28%
18	33 kW	37%	28%
19	34 kW	36%	28%
20	35 kW	35%	28%
21	36 kW	34%	26%
22	37 kW	33%	26%
23	38 kW	32%	26%
24	39 kW	31%	26%
25	40 kW	30%	26%
26-30	15 kW plus 1 kW for each range	30%	24%
31-40		30%	22%
41-50	25 kW plus ¾ kW for each range	30%	20%
51-60		30%	18%
61 & over		30%	16%

Note 1. Over 12 kW to 27 kW ranges all of same rating. For ranges, individually rated more than 12 kW but not more than 27 kW, the maximum demand in Column A shall be increased 5 percent for each additional kW of rating or major fraction thereof by which the rating of individual ranges exceeds 12 kW.

Note 2. Over 12 kW to 27 kW ranges *of unequal ratings*. For ranges individually rated more than 12 kW and of different ratings but none exceeding 27 kW an average value of rating shall be calculated by adding together the ratings of all ranges to obtain the total connected load (using 12 kW for any range rated less than 12 kW) and dividing by the total number of ranges; and then the maximum demand in Column A

Useful Information

shall be increased 5 percent for each kW or major fraction thereof by which this average value exceeds 12 kW.

Note 3. This table does not apply to commercial ranges. See Table 220-6(a) for demand factors for commercial cooking equipment.

Note 4. Over 1¾ kW to 8¾ kW. In lieu of the method provided in Column A, loads rated more than 1¾ kW but not more than 8¾ kW may be considered as the sum of the nameplate ratings of all the loads, multiplied by the demand factors specified in Columns B or C for the given number of loads.

Note 5. Branch-Circuit Load. Branch-circuit load for one range may be computed in accordance with Table 220-5. The branch-circuit load for one wall-mounted oven or one counter-mounted cooking unit shall be the nameplate rating of the appliance. The branch-circuit load for a counter-mounted cooking unit and not more than two wall-mounted ovens, all supplied from a single branch circuit and located in the same room shall be computed by adding the nameplate ratings of the individual appliances and treating this total as equivalent to one range.

Table 250-94(a)

Grounding Electrode Conductor for Grounded Systems

Size of Largest Service-Entrance Conductor or Equivalent for Parallel Conductors		Size of Grounding Electrode Conductor	
Copper	Aluminum or Copper-Clad Aluminum	Copper	*Aluminum or Copper-Clad Aluminum
2 or smaller	0 or smaller	8	6
1 or 0	2/0 or 3/0	6	4
2/0 or 3/0	4/0 or 250 MCM	4	2
Over 3/0 thru 350 MCM	Over 250 MCM thru 500 MCM	2	0
Over 350 MCM thru 600 MCM	Over 500 MCM thru 900 MCM	0	3/0
Over 600 MCM thru 1100 MCM	Over 900 MCM thru 1750 MCM	2/0	4/0
Over 1100 MCM	Over 1750 MCM	3/0	250 MCM

Where there are no service-entrance conductors, the grounding electrode conductor size shall be determined by the equivalent size of the largest service-entrance conductor required for the load to be served.

* See installation restrictions in Section 250-92(a).

See Section 250-23(b).

House Wiring Simplified

Table 310-2(a). Conductor Application

Trade Name	Type Letter	Max. Operating Temp.	Application Provisions
Rubber-Covered Fixture Wire Solid or 7-Strand	*RF-1	60°C 140°F	Fixture wiring. Limited to 300 volts.
	*RF-2	60°C 140°F	Fixture wiring, and as permitted in Section 725-14.
Rubbered-Covered Fixture Wire Flexible Stranding	*FF-1	60°C 140°F	Fixture wiring. Limited to 300 volts.
	*FF-2	60°C 140°F	Fixture wiring, and as permitted in Section 725-14.
Heat-Resistant Rubber-Covered Fixture Wire Solid or 7-Strand	*RFH-1	75°C 167°F	Fixture wiring. Limited to 300 volts.
	*RFH-2	75°C 167°F	Fixture wiring, and as permitted in Section 725-14.
Heat-Resistant Rubber-Covered Fixture Wire Flexible Stranding	*FFH-1	75°C 167°F	Fixture wiring. Limited to 300 volts.
	*FFH-2	75°C 167°F	Fixture wiring, and as permitted in Section 725-14.
Thermoplastic-Covered Fixture Wire—Solid or Stranded	*TF	60°C 140°F	Fixture wiring, and as permitted in Section 725-14.
Thermoplastic-Covered Fixture Wire—Flexible Stranding	*TFF	60°C 140°F	Fixture wiring, and as permitted in Section 725-14.
Heat Resistant, Thermoplastic—Covered Fixture Wire—Solid or Stranded	*TFN	90°C	Fixture wiring, and as permitted in Section 725-14.
Heat Resistant Thermoplastic—Covered Fixture Wire—Flexible Stranding	*TFFN	90°C	Fixture wiring, and as permitted in Section 725-14.
Cotton-Covered, Heat-Resistant, Fixture Wire	*CF	90°C 194°F	Fixture wiring. Limited to 300 volts.

* Fixture wires are not intended for installation as branch-circuit conductors except as permitted in Section 725-14.

Useful Information

Table 310-2(a)—Continued

Trade Name	Type Letter	Max. Operating Temp.	Application Provisions
Asbestos-Covered Heat-Resistant, Fixture Wire	*AF	150°C 302°F	Fixture wiring. Limited to 300 volts. and Indoor Dry Location.
Fluorinated Ethylene Propylene Fixture Wire Solid or 7 Strand	*PF *PGF	200°C 392°F	Fixture Wiring and as permitted in Section 725-14.
Fluorinated Ethylene Propylene Fixture Wire	*PFF *PGFF	150°C 302°F	Fixture Wiring and as permitted in Section 725-14.
Extruded Polytetra-fluoroethylene (PTFE) Solid or 7-Strand	*PTF	250°C 482°F	Fixture wire, and as permitted in Section 725-14. (Nickel or nickel-coated copper)
Extruded Polytetra-fluoroethylene (PTFE) Flexible Stranding (#26-#36 AWG)	*PTFF	150°C 302°F	Fixture wire, and as permitted in Section 725-14. (Silver or nickel-coated copper)
Silicone Rubber Insulated Fixture Wire	*SF-1	200°C 392°F	Fixture wiring. Limited to 300 volts.
Solid or 7 Strand	*SF-2	200°C 392°F	Fixture wiring and as permitted in Section 725-14.
Silicone Rubber Insulated Fixture Wire	*SFF-1	150°C 302°F	Fixture wiring. Limited to 300 volts.
Flexible Stranding	*SFF-2	150°C 302°F	Fixture wiring and as permitted in Section 725-14.
Heat-Resistant Rubber	RH	75°C 167°F	Dry locations.
Heat-Resistant Rubber	RHH	90°C 194°F	Dry locations.

* Fixture wires are not intended for installation as branch-circuit conductors except as permitted in Section 725-14.

House Wiring Simplified

Table 310-2(a)—Continued

Trade Name	Type Letter	Max. Operating Temp.	Application Provisions
Moisture and Heat-Resistant Rubber	RHW	75°C 167°F	Dry and wet locations. For over 2,000 volts, insulation shall be ozone-resistant.
Heat-Resistant Latex Rubber	RUH	75°C	Dry locations.
Moisture Resistant Latex Rubber	RUW	60°C 140°F	Dry and wet locations.
Thermoplastic	T	60°C 140°F	Dry locations.
Moisture-Resistant Thermoplastic	TW	60°C 140°F	Dry and wet locations.
Heat-Resistant Thermoplastic	THHN	90°C 194°F	Dry locations.
Moisture and Heat-Resistant Thermoplastic	THW	75°C 167°F	Dry and wet locations.
		90°C 194°F	Special applications *within* electric discharge lighting equipment. Limited to 1000 open-circuit volts or less. (Size 14-8 only as permitted in Section 410-26.)
Moisture and Heat-Resistant Thermoplastic	THWN	75°C 167°F	Dry and wet locations.
Moisture and Heat-Resistant Cross-Linked Synthetic Polymer	XHHW	90°C 194°F 75°C 167°F	Dry locations. Wet locations.
Moisture-, Heat- and Oil-Resistant Thermoplastic	MTW	60°C 140°F	Machine Tool Wiring in wet locations as permitted in NFPA Standard No. 79 (See Article 670).
		90°C 194°F	Machine Tool Wiring in dry locations as permitted in NFPA Standard No. 79 (See Article 670).
Thermoplastic and Asbestos	TA	90°C 194°F	Switchboard wiring only.

Useful Information

Table 310-2(a)—Continued

Trade Name	Type Letter	Max. Operating Temp.	Application Provisions
Thermoplastic and Fibrous Outer Braid	TBS	90°C 194°F	Switchboard wiring only.
Synthetic Heat-Resistant	SIS	90°C 194°F	Switchboard wiring only.
Mineral Insulation (Metal Sheathed)	MI	85°C 185°F	Dry and wet locations.
		250°C 482°F	For special application.
Extruded Polytetrafluoroethylene	TFE	250°C 482°F	Dry locations only. Only for leads within apparatus or within raceways connected to apparatus, or as open wiring. (Nickel or nickel-coated copper only.)
Silicone-Asbestos	SA	90°C 194°F	Dry locations.
		125°C 257°F	For special application.
Fluorinated Ethylene Propylene	FEP or FEPB	90°C 194°F 200°C 392°F	Dry locations. Dry locations—special applications.
Varnished Cambric	V	85°C 185°F	Dry locations only. Smaller than No. 6 by special permission.
Asbestos and Varnished Cambric	AVA	110°C 230°F	Dry locations only.
Asbestos and Varnished Cambric	AVL	110°C 230°F	Dry and wet locations.
Asbestos and Varnished Cambric	AVB	90°C 194°F	Dry locations only.
Asbestos	A	200°C 392°F	Dry locations only. Only for leads within apparatus or within raceways connected to apparatus. Limited to 300 volts.

Table 310-2(a)—Continued

Trade Name	Type Letter	Max. Operating Temp.	Application Provisions
Asbestos	AA	200°C 392°F	Dry locations only. Only for leads within apparatus or within raceways connected to apparatus or as open wiring. Limited to 300 volts.
Asbestos	AI	125°C 257°F	Dry locations only. Only for leads within apparatus or within raceways connected to apparatus. Limited to 300 volts.
Asbestos	AIA	125°C 257°F	Dry locations only. Only for leads within apparatus or within raceways connected to apparatus or as open wiring.
Paper		85°C 185°F	For underground service conductors, or by special permission.

Table 310-12.   Allowable Ampacities of Insulated Copper Conductors

Not More than Three Conductors in Raceway or Cable or
Direct Burial (Based on Ambient Temperature of 30° C. 86°F.)

Size	Temperature Rating of Conductor. See Table 310-2(a)							
AWG MCM	60°C (140°F)	75°C (167°F)	85°C (185°F)	90°C (194°F)	110°C (230°F)	125°C (257°F)	200°C (392°F)	250°C (482°F)
	TYPES RUW (14-2), T, TW	TYPES RH, RHW, RUH (14-2), THW, THWN, XHHW	TYPES V, MI	TYPES TA, TBS, SA, AVB, SIS, FEP, FEPB, RHH, THHN, XHHW**	TYPES AVA, AVL	TYPES AI (14-8), AIA	TYPES A (14-8), AA, FEP*, FEPB*	TYPE TFE (Nickel or nickel-coated copper only)
14	15	15	25	25†	30	30	30	40
12	20	20	30	30†	35	40	40	55
10	30	30	40	40†	45	50	55	75
8	40	45	50	50	60	65	70	95
6	55	65	70	70	80	85	95	120
***4	70	85	90	90	105	115	120	145
***3	80	100	105	105	120	130	145	170
***2	95	115	120	120	135	145	165	195
***1	110	130	140	140	160	170	190	220
***0	125	150	155	155	190	200	225	250
***00	145	175	185	185	215	230	250	280
000	165	200	210	210	245	265	285	315
0000	195	230	235	235	275	310	340	370
250	215	255	270	270	315	335	....	....
300	240	285	300	300	345	380	....	....
350	260	310	325	325	390	420	....	....
400	280	335	360	360	420	450	....	....
500	320	380	405	405	470	500	....	....
600	355	420	455	455	525	545	....	....
700	385	460	490	490	560	600	....	....
750	400	475	500	500	580	620	....	....
800	410	490	515	515	600	640	....	....
900	435	520	555	555	....	....	....	....
1000	455	545	585	585	680	730	....	....
1250	495	590	645	645	....	....	....	....
1500	520	625	700	700	785	....	....	....
1750	545	650	735	735	....	....	....	....
2000	560	665	775	775	840	....	....	....

* Special use only. See Table 310-2(a).
** For dry locations only. See Table 310-2(a).
These ampacities relate only to conductors described in Table 310-2(a).
*** For 3-wire, single-phase residential services, the allowable ampacity of RH, RHH, RHW, THW and XHHW copper conductors shall be for sizes No. 4-100 Amp., No. 3-110 Amp., No. 2-125 Amp., No. 1-150 Amp., No. 1/0-175 Amp., and No. 2/0-200 Amp.
† The ampacities for Types FEP, FEPB, RHH, THHN, and XHHW conductors for sizes AWG 14, 12 and 10 shall be the same as designated for 75°C conductors in this Table.
For ambient temperatures over 30°C, see Correction Factors, Note 13.

NATIONAL ELECTRICAL CODE

Table 310-13. Allowable Ampacities of Insulated Copper Conductors

Single Conductor in Free Air
(Based on Ambient Temperature of 30°C. 86°F.)

Size	Temperature Rating of Conductor. See Table 310-2(a)								
AWG MCM	60°C (140°F)	75°C (167°F)	85°C (185°F)	90°C (194°F)	110°C (230°F)	125°C (257°F)	200°C (392°F)	250°C (482°F)	
	TYPES RUW (14-2), T, TW	TYPES RH, RHW, RUH (14-2), THW, THWN, XHHW	TYPES V, MI	TYPES TA, TBS, SA, AVB, SIS, FEP, FEPB, RHH, THHN, XHHW**	TYPES AVA, AVL	TYPES AI (14-8), AIA	TYPES A (14-8), AA, FEP* FEPB*	TYPE TFE (Nickel or nickel-coated copper only)	Bare and Covered Conductors
14	20	20	30	30†	40	40	45	60	30
12	25	25	40	40†	50	50	55	80	40
10	40	40	55	55†	65	70	75	110	55
8	55	65	70	70	85	90	100	145	70
6	80	95	100	100	120	125	135	210	100
4	105	125	135	135	160	170	180	285	130
3	120	145	155	155	180	195	210	335	150
2	140	170	180	180	210	225	240	390	175
1	165	195	210	210	245	265	280	450	205
0	195	230	245	245	285	305	325	545	235
00	225	265	285	285	330	355	370	605	275
000	260	310	330	330	385	410	430	725	320
0000	300	360	385	385	445	475	510	850	370
250	340	405	425	425	495	530	....	....	410
300	375	445	480	480	555	590	....	....	460
350	420	505	530	530	610	655	....	....	510
400	455	545	575	575	665	710	....	....	555
500	515	620	660	660	765	815	....	....	630
600	575	690	740	740	855	910	....	....	710
700	630	755	815	815	940	1005	....	....	780
750	655	785	845	845	980	1045	....	....	810
800	680	815	880	880	1020	1085	....	....	845
900	730	870	940	940	....	....	....	....	905
1000	780	935	1000	1000	1165	1240	....	....	965
1250	890	1065	1130	1130	....	....	....	....	....
1500	980	1175	1260	1260	1450	....	....	....	1215
1750	1070	1280	1370	1370	....	....	....	....	....
2000	1155	1385	1470	1470	1715	....	....	....	1405

* Special use only. See Table 310-2(a).
** For dry locations only. See Table 310-2(a).
These ampacities relate only to conductors described in Table 310-2(a).
† The ampacities for Types FEP, FEPB, RHH, THHN, and XHHW conductors for sizes AWG 14, 12 and 10 shall be the same as designated for 75°C conductors in this Table.
For ambient temperatures over 30°C, see Correction Factors, Note 13.

Useful Information

Table 310-14. Allowable Ampacities of Insulated Aluminum and Copper-Clad Aluminum Conductors

Not More than Three Conductors in Raceway or Cable or Direct Burial (Based on Ambient Temperature of 30°C. 86°F.)

Size	Temperature Rating of Conductor. See Table 310-2(a)						
AWG MCM	60°C (140°F)	75°C (167°F)	85°C (185°F)	90°C (194°F)	110°C (230°F)	125°C (257°F)	200°C (392°F)
	TYPES RUW (12-2), T, TW	TYPES RH, RHW, RUH (12-2), THW, THWN XHHW	TYPES V, MI	TYPES TA, TBS, SA, AVB, SIS, RHH THHN XHHW**	TYPES AVA, AVL	TYPES AI (12-8), AIA	TYPES A (12-8), AA
12	15	15	25	25 †	25	30	30
10	25	25	30	30 †	35	40	45
8	30	40	40	40	45	50	55
6	40	50	55	55	60	65	75
4	55	65	70	70	80	90	95
3	65	75	80	80	95	100	115
*2	75	90	95	95	105	115	130
*1	85	100	110	110	125	135	150
*0	100	120	125	125	150	160	180
*00	115	135	145	145	170	180	200
*000	130	155	165	165	195	210	225
*0000	155	180	185	185	215	245	270
250	170	205	215	215	250	270	....
300	190	230	240	240	275	305	....
350	210	250	260	260	310	335	....
400	225	270	290	290	335	360	....
500	260	310	330	330	380	405	....
600	285	340	370	370	425	440	....
700	310	375	395	395	455	485	....
750	320	385	405	405	470	500	....
800	330	395	415	415	485	520	....
900	355	425	455	455	....	....	....
1000	375	445	480	480	560	600	....
1250	405	485	530	530	....	....	....
1500	435	520	580	580	650	....	....
1750	455	545	615	615	....	....	....
2000	470	560	650	650	705	....	....

These ampacities relate only to conductors described in Table 310-2(a).

* For 3-wire, single-phase residential services, the allowable ampacity of RH, RHH, RHW, THW, and XHHW conductors shall be for sizes No. 2-100 Amp., No. 1-110 Amp., No. 1/0-125 Amp., No. 2/0-150 Amp., No. 3/0-175 Amp. and No. 4/0-200 Amp.

** For dry locations only. See Table 310-2(a).

† The ampacities for Type RHH, THHN, and XHHW conductors for sizes AWG 12 and 10 shall be the same as designated for 75°C conductors in this Table.

For ambient temperatures over 30°C, see Correction Factors, Note 13.

153

House Wiring Simplified

Table 310-15.   Allowable Ampacities of Insulated Aluminum and Copper-Clad Aluminum Conductors

Single Conductor in Free Air
(Based on Ambient Temperature of 30°C. 86°F.)

Size	Temperature Rating of Conductor. See Table 310-2(a)							
AWG MCM	60°C (140°F)	75°C (167°F)	85°C (185°F)	90°C (194°F)	110°C (230°F)	125°C (257°F)	200°C (392°F)	
	TYPES RUW (12-2), T, TW	TYPES RH, RHW, RUH (12-2), THW, THWN, XHHW	TYPES V, MI	TYPES TA, TBS, SA, AVB, SIS, RHH, THHN, XHHW*	TYPES AVA, AVL	TYPES AI (12-8), AIA	TYPES A (12-8), AA	Bare and Covered Conductors
12	20	20	30	30 †	40	40	45	30
10	30	30	45	45 †	50	55	60	45
8	45	55	55	55	65	70	80	55
6	60	75	80	80	95	100	105	80
4	80	100	105	105	125	135	140	100
3	95	115	120	120	140	150	165	115
2	110	135	140	140	165	175	185	135
1	130	155	165	165	190	205	220	160
0	150	180	190	190	220	240	255	185
00	175	210	220	220	255	275	290	215
000	200	240	255	255	300	320	335	250
0000	230	280	300	300	345	370	400	290
250	265	315	330	330	385	415	....	320
300	290	350	375	375	435	460	....	360
350	330	395	415	415	475	510	....	400
400	355	425	450	450	520	555	....	435
500	405	485	515	515	595	635	....	490
600	455	545	585	585	675	720	....	560
700	500	595	645	645	745	795	....	615
750	515	620	670	670	775	825	....	640
800	535	645	695	695	805	855	....	670
900	580	700	750	750	....	....	....	725
1000	625	750	800	800	930	990	....	770
1250	710	855	905	905	....	....	....	....
1500	795	950	1020	1020	1175	....	....	985
1750	875	1050	1125	1125	....	....	....	....
2000	960	1150	1220	1220	1425	....	....	1165

These ampacities relate only to conductors described in Table 310-2(a).

* For dry locations only. See Table 310-2(a).

† The ampacities for Types RHH, THHN, and XHHW conductors for sizes AWG 12 and 10 shall be the same as designated for 75°C conductors in this Table.

For ambient temperatures over 30°C, see Correction Factors, Note 13.

Table 310-21. Simplified Wiring Table (See Section 310-20 for use)
Conductor Size*—6 or Fewer Conductors in Raceway or Cable

Am-peres	Copper				Aluminum and Copper-Clad Aluminum			
	Non-Cont.		Continuous		Non-Cont.		Continuous	
	AWG	MCM	AWG	MCM	AWG	MCM	AWG	MCM
15	14		14		12		12	
20	12		12		10		10	
25	10		10		8		8	
30	10		10		8		8	
35	8		8		6		6	
40	8		8		6		6	
45	6		6		4		4	
50	6		6		4		4	
60	4		4		4		4	
70	4		4		3		3	
80	3		3		3		2	
90	3		2		2		1	
100	2		1		1		0	
110	1		0		0		2/0	
125	1		0		2/0		3/0	
150	0		2/0		3/0		4/0	
175	2/0		3/0		4/0			250
200	3/0		4/0			250		300
225	4/0			250		300		350
250		250		300		350		400
300		350		400		400		750
350		400		500		500		1000
400		500		750		750		
450		750		1000		1000		
500		750				1000		
600		1000						

* Neutral conductors shall be treated in accordance with Note 10 — Neutral Conductors of Notes to Tables 310-12 through 310-15.

346-10. Bends — How Made. Bends of rigid conduit shall be so made that the conduit will not be injured, and that the internal diameter of the conduit will not be effectively reduced. The radius of the curve of the inner edge of any field bend shall not be less than shown in Table 346-10(a).

Exception: For field bends for conductors without lead sheath and made with a single operation (one shot) bending machine designed for the purpose, the minimum radius may be in accordance with Table 346-10(b).

Table 346-10(a)

Radius of Conduit Bends (Inches)

Size of Conduit (In.)	Conductors Without Lead Sheath (In.)	Conductors With Lead Sheath (In.)
½	4	6
¾	5	8
1	6	11
1¼	8	14
1½	10	16
2	12	21
2½	15	25
3	18	31
3½	21	36
4	24	40
4½	27	45
5	30	50
6	36	61

Table 346-10(b)

Radius of Conduit Bends (Inches)

Size of Conduit (In.)	Radius to Center of Conduit (In.)
½	4
¾	4½
1	5¾
1¼	7¼
1½	8¼
2	9½
2½	10½
3	13
3½	15
4	16
4½	20
5	24
6	30

Useful Information

Table 346-12.   Supports for Rigid Metal Conduit

Conduit Size (Inches)	Maximum distance between rigid metal conduit supports (Feet)
½–¾	10
1	12
1¼–1½	14
2–2½	16
3 and larger	20

Table 350-3. Maximum Number of Insulated Conductors in ⅜ In. Flexible Metal Conduit.*

Col. A = With fitting inside conduit.
Col. B = With fitting outside conduit.

Size AWG	Types RF-2, RFH-2, SF-2		Types TF, T, XHHW, AF, TW, RUH, RUW		Types TFN, THHN, THWN		Types FEP, FEPB, PF, PGF	
	A	B	A	B	A	B	A	B
18	..	3	3	7	4	8	5	8
16	..	2	2	4	3	7	4	8
14	..	..	..	4	3	7	3	7
12	..	..	..	3	..	4	..	4
10	..	..	..	..	..	2	..	3

* In addition one uninsulated grounding conductor of the same AWG size may be installed.

Table 400-9(b). Ampacity of Flexible Cord

Table 400-9(b) gives the allowable ampacity for not more than 3 current-carrying conductors in a cord. If the number of current-carrying conductors in a cord is from 4 to 6, the allowable ampacity of each conductor shall be reduced to 80 percent of the values for not more than 3 current-carrying conductors in the Table. A conductor used for equipment grounding and a neutral conductor which carries only the unbalanced current from other conductors, as in the case of normally balanced circuits of 3 or more conductors, are not considered to be current-carrying conductors. Where a single conductor is used for both equipment grounding and to carry unbalanced current from other conductors, it shall not be considered to be a current-carrying conductor. (See Section 250-60.)

(Based on Ambient Temperature of 30°C (86°F). See Section 400-9 and Table 400-11)

Size AWG	Rubber Types TP, TS / Thermoplastic Types TPT, TST	Rubber Types PO, C, PD, E, EO, EN, S, SO, SRD, SJ, SJO, SV, SVO, SP / Thermoplastic Types ET, ETT, ETLB, ETP, ST, STO, SRDT, SJT, SJTO, SVT, SVTO, SPT		Types AFS, AFSJ, HC, HPD, HSJ, HSJO, HS, HSO, HPN, SVHT	Types AVPO AVPD	Cotton Types CFPD* / Asbestos Types AFC* AFPD*
		A†	B†			
27**	0.5	..	..	..	..	..
18	..	7	10	10	17	6
17	..	..	..	12	..	..
16	..	10	13	15	22	8
15	..	..	..	17	..	..
14	..	15	18	20	28	17
12	..	20	25	30	36	23
10	..	25	30	35	47	28
8	..	35	40	..	..	..
6	..	45	55	..	..	..
4	..	60	70	..	..	..
2	..	80	95	..	..	..

* These types are used almost exclusively in fixtures where they are exposed to high temperatures and ampere ratings are assigned accordingly.

** Tinsel Cord.

† The ampacities under sub-heading A are applicable to 3-conductor cords and other multi-conductor cords connected to utilization equipment so that only 3 conductors are current carrying. The ampacities under sub-heading B are applicable to 2-conductor cords and other multi-conductor cords connected to utilization equipment so that only 2 conductors are current carrying.

NOTE 1. Ultimate Insulation Temperature. In no case shall conductors be associated together in such way with respect to the kind of circuit, the wiring method employed, or the number of conductors, that the limiting temperature of the conductors will be exceeded.

NOTE 2. SVHT made only in No. 18 and 17 AWG sizes.

Table 2
Maximum Number of Fixture Wires in Trade Size of Conduit or Tubing
(40 Percent Fill Based on Individual Diameters)

Conduit Trade Size (Inches)	½			¾			1			1¼			1½			2		
Wire Types	18	16	14	18	16	14	18	16	14	18	16	14	18	16	14	18	16	14
PTF, PTFF, PGFF, PGF, PFF, PF	23	18	14	40	31	24	65	50	39	115	90	70	157	122	95	257	200	156
TFFN, TFN	19	15		34	26		55	43		97	76		132	104		216	169	
SF-1	16			29			47			83			114			186		
SFF-1, FF-1, FFH-1	15			26			43			76			104			169		
CF	13	10	8	23	18	14	38	30	23	66	53	40	91	72	55	149	118	90
TF	11	10		20	18		32	30		57	53		79	72		129	118	
RFH-1, RF-1	11			20			32			57			79			129		
TFF	11	10		20	17		32	27		56	49		77	66		126	109	
AF	11	9	7	19	16	12	31	26	20	55	46	36	75	63	49	123	104	81
SFF-2	9	7	6	16	12	10	27	20	17	47	36	30	65	49	42	106	81	68
SF-2	9	8	6	16	14	11	27	23	18	47	40	32	65	55	43	106	90	71
FF-2, FFH-2	9	7		15	12		25	19		44	34		60	46		99	75	
RFH-2	7	5		12	10		20	16		36	28		49	38		80	62	
RF-2	7	6		12	10		20	16		36	29		49	40		80	65	

Table 3A
Maximum Number of Conductors in Trade Sizes of Conduit or Tubing
(Based on Table 1, Chapter 9)

Type Letters	Conductor Size AWG, MCM	½	¾	1	1¼	1½	2	2½	3	3½	4	4½	5	6
TW, T, RUH, RUW, XHHW (14 thru 8)	14	9	15	25	44	60	99	142	171	176	124			
	12	7	12	19	35	47	78	111	131	97				
	10	5	9	15	26	36	60	85	72					
	8	3	5	8	14	20	33	47						
RHW and RHH (without outer covering), THW	14	6	10	16	29	40	65	93	143	192	163	121	152	
	12	4	8	13	24	32	53	76	117	157	96			
	10	4	6	11	19	26	43	61	95	127				
	8	1	4	6	11	15	25	36	56	75				
TW, T, THW, RUH (6 thru 2), RUW (6 thru 2)	6	1	2	4	7	10	16	23	36	48	62	78	97	141
	4	1	1	3	5	7	12	17	27	36	47	58	73	106
	3	1	1	2	4	6	10	15	23	31	40	50	63	91
	2	1	1	2	4	5	9	13	20	27	34	43	54	78
	1		1	1	3	4	6	9	14	19	25	31	39	57
FEPB (6 thru 2), RHW and RHH (without outer covering)	0		1	1	2	3	5	8	12	16	21	27	33	49
	00		1	1	1	3	5	7	10	14	18	23	29	41
	000		1	1	1	2	4	6	9	12	15	19	24	35
	0000			1	1	1	3	5	7	10	13	16	20	29
	250				1	1	2	4	6	8	10	13	16	23
	300				1	1	2	3	5	7	9	11	14	20
	350				1	1	1	3	4	6	8	10	12	18
	400				1	1	1	2	4	5	7	9	11	16
	500					1	1	1	3	4	6	7	9	14
	600				1	1	1	1	3	4	5	6	7	11
	700					1	1	1	2	3	4	5	7	10
	750					1	1	1	2	3	4	5	6	9

Table 3B
Maximum Number of Conductors in Trade Sizes of Conduit or Tubing
(Based on Table 1, Chapter 9)

Type Letters	Conductor Size AWG, MCM	½	¾	1	1¼	1½	2	2½	3	3½	4	4½	5	6
THWN, THHN,	14	13	24	39	69	94	154							
	12	10	18	29	51	70	114	164						
	10	6	11	18	32	44	73	104	160					
	8	3	6	10	19	26	42	60	93	125	160			
THHN, FEP (14 thru 2), FEPB (14 thru 8),	6	1	4	6	11	15	26	37	57	76	98	125	154	
	4	1	2	4	7	9	16	22	35	47	60	75	94	137
	3	1	1	3	6	8	13	19	29	39	51	64	80	116
	2	1	1	3	5	7	11	16	25	33	43	54	67	97
	1		1	1	3	5	8	12	18	25	32	40	50	72
XHHW (4 thru 500MCM)	0		1	1	3	4	7	10	15	21	27	33	42	61
	00		1	1	2	3	6	8	13	17	22	28	35	51
	000		1	1	1	3	5	7	11	14	18	23	29	42
	0000		1	1	1	2	4	6	9	12	15	19	24	35
	250			1	1	1	3	4	7	10	12	16	20	28
	300			1	1	1	3	4	6	8	11	13	17	24
	350			1	1	1	2	3	5	7	9	12	15	21
	400				1	1	1	3	5	6	8	10	13	19
	500				1	1	1	2	4	5	7	9	11	16
	600				1	1	1	1	3	4	5	7	9	13
	700					1	1	1	3	4	5	6	8	11
	750					1	1	1	2	3	4	6	7	11
XHHW	6	1	3	5	9	13	21	30	47	63	81	102	128	185
	600				1	1	1	1	3	4	5	7	9	13
	700					1	1	1	3	4	5	6	7	11
	750					1	1	1	2	3	4	6	7	10

Table 3C
Maximum Number of Conductors in Trade Sizes of Conduit or Tubing
(Based on Table 1, Chapter 9)

Type Letters	Conductor Size AWG, MCM	½	¾	1	1¼	1½	2	2½	3	3½	4	4½	5	6
RHW.	14	3	6	10	18	25	41	58	90	121	155			
	12	3	5	9	15	21	35	50	77	103	132			
	10	2	4	7	13	18	29	41	64	86	110	138		
	8	1	2	4	8	10	17	25	39	52	67	84	105	152
RHH (with outer covering)	6	1	1	2	5	6	11	15	24	32	41	51	64	93
	4	1	1	1	3	5	8	12	18	24	31	39	50	72
	3	1	1	1	3	4	7	10	16	22	28	35	44	63
	2		1	1	3	4	6	9	14	19	24	31	38	56
	1			1	1	3	5	7	11	14	18	23	29	42
	0		1	1	1	2	4	6	9	12	16	20	25	37
	00			1	1	1	3	5	8	11	14	18	22	32
	000			1	1	1	3	4	7	9	12	15	19	28
	0000			1	1	1	2	4	6	8	10	13	16	24
	250				1	1	1	3	5	6	8	11	13	19
	300				1	1	1	3	4	5	7	9	11	17
	350				1	1	1	2	4	5	6	8	10	15
	400							1	3	4	6	7	9	14
	500				1	1	1	1	3	4	5	6	8	11
	600					1	1	1	2	3	4	5	6	9
	700						1	1	1	3	3	4	6	8
	750							1	1	3	3	4	5	8

Table 8. Properties of Conductors

Size AWG MCM	Area Cir. Mils	Concentric Lay Stranded Conductors		Bare Conductors		D. C. Resistance Ohms/M Ft. At 25°C. 77°F.		
						Copper		Alumni-num
		No. Wires	Diam. Each Wire Inches	Diam. Inches	*Area Sq. Inches	Bare Cond.	Tin'd. Cond.	
18	1620	Solid	.0403	.0403	.0013	6.51	6.79	10.7
16	2580	Solid	.0508	.0508	.0020	4.10	4.26	6.72
14	4110	Solid	.0641	.0641	.0032	2.57	2.68	4.22
12	6530	Solid	.0808	.0808	.0051	1.62	1.68	2.66
10	10380	Solid	.1019	.1019	.0081	1.018	1.06	1.67
8	16510	Solid	.1285	.1285	.0130	.6404	.659	1.05
6	26240	7	.0612	.184	.027	.410	.427	.674
4	41740	7	.0772	.232	.042	.259	.269	.424
3	52620	7	.0867	.260	.053	.205	.213	.336
2	66360	7	.0974	.292	.067	.162	.169	.266
1	83690	19	.0664	.332	.087	.129	.134	.211
0	105600	19	.0745	.372	.109	.102	.106	.168
00	133100	19	.0837	.418	.137	.0811	.0843	.133
000	167800	19	.0940	.470	.173	.0642	.0668	.105
0000	211600	19	.1055	.528	.219	.0509	.0525	.0836
250	250000	37	.0822	.575	.260	.0431	.0449	.0708
300	300000	37	.0900	.630	.312	.0360	.0374	.0590
350	350000	37	.0973	.681	.364	.0308	.0320	.0505
400	400000	37	.1040	.728	.416	.0270	.0278	.0442
500	500000	37	.1162	.813	.519	.0216	.0222	.0354
600	600000	61	.0992	.893	.626	.0180	.0187	.0295
700	700000	61	.1071	.964	.730	.0154	.0159	.0253
750	750000	61	.1109	.998	.782	.0144	.0148	.0236
800	800000	61	.1145	1.030	.833	.0135	.0139	.0221
900	900000	61	.1215	1.090	.933	.0120	.0123	.0197
1000	1000000	61	.1280	1.150	1.039	.0108	.0111	.0177
1250	1250000	91	.1172	1.289	1.305	.00863	.00888	.0142
1500	1500000	91	.1284	1.410	1.561	.00719	.00740	.0118
1750	1750000	127	.1174	1.526	1.829	.00616	.00634	.0101
2000	2000000	127	.1255	1.630	2.087	.00539	.00555	.00885

* Area given is that of a circle having a diameter equal to the over-all diameter of a stranded conductor.

The values given in the Table are those given in Handbook 100 of the National Bureau of Standards except that those shown in the 8th column are those given in Specification B33 of the American Society for Testing and Materials, and those shown in the 9th column are those given in Standard No. S-19-81 of the Insulated Power Cable Engineers Association and Standard No. WC3-1964 of the National Electrical Manufacturers Association.

The resistance values given in the last three columns are applicable only to direct current. When conductors larger than No. 4/0 are used with alternating current the multiplying factors in Table 9, Chapter 9 should be used to compensate for skin effect.

House Wiring Simplified

Table 9. Multiplying Factors for Converting DC Resistance to 60-Hertz AC Resistance

Size		Multiplying Factor			
		For Nonmetallic Sheathed Cables in Air or Nonmetallic Conduit		For Metallic Sheathed Cables or all Cables in Metallic Raceways	
		Copper	Aluminum	Copper	Aluminum
Up to	3 AWG	1.	1.	1.	1.
	2	1.	1.	1.01	1.00
	1	1.	1.	1.01	1.00
	0	1.001	1.000	1.02	1.00
	00	1.001	1.001	1.03	1.00
	000	1.002	1.001	1.04	1.01
	0000	1.004	1.002	1.05	1.01
	250 MCM	1.005	1.002	1.06	1.02
	300 MCM	1.006	1.003	1.07	1.02
	350 MCM	1.009	1.004	1.08	1.03
	400 MCM	1.011	1.005	1.10	1.04
	500 MCM	1.018	1.007	1.13	1.06
	600 MCM	1.025	1.010	1.16	1.08
	700 MCM	1.034	1.013	1.19	1.11
	750 MCM	1.039	1.015	1.21	1.12
	800 MCM	1.044	1.017	1.22	1.14
	1000 MCM	1.067	1.026	1.30	1.19
	1250 MCM	1.102	1.040	1.41	1.27
	1500 MCM	1.142	1.058	1.53	1.36
	1750 MCM	1.185	1.079	1.67	1.46
	2000 MCM	1.233	1.100	1.82	1.56

DICTIONARY OF TERMS

ACCESSIBLE: Easy to get to. An electrical outlet is said to be accessible if it may be worked on without disturbing the finish or construction of a house.

ALTERNATING CURRENT: In alternating current, the voltage flows in one direction one instant, and the other direction the next instant. The direction of flow reverses continually. Each two reversals is called a cycle. The number of cycles per second is called frequency. In the United States, most systems are 60 cycles.

AMMETER: Instrument used to measure amount of electric current flow in amperes.

AMPACITY: Current-carrying capacity of conductor expressed in amperes.

AMPERE: Measure of the flow of electricity. In measuring the amount of water passing through a pipe, the term gallons is used. In electricity, the term ampere is the unit of measurement. Various parts of a wiring system . . . fuses, wall switches, fuse boxes, etc. are rated in amperes. Ratings in amperes indicate the greatest amount of current with which these parts should be used.

APPLIANCE; FIXED: Appliance fastened or otherwise secured at specific location.

APPLIANCE; PORTABLE: Appliance which may be easily moved from one place to another in normal use.

APPLIANCE; STATIONARY: An appliance which is not easily moved from one place to another in normal use.

APPROVED: Signifies that minimum standards established by an authority have been met.

ARC, ELECTRIC: Sustained visible discharge of electricity across gap in circuit or between electrodes. Arcing takes place in switches and other make-and-break devices when a circuit is opened, and at the brushes of a commutator type motor if brush contact is bad.

ARMORED CABLE: A flexible, metallic sheathed cable used for interior wiring. Commonly called BX.

ATOM: Smallest particle that makes up type of matter called an element. The element retains its characteristics when subdivided into atoms. Examples of elements are hydrogen, oxygen, helium. Over 100 elements have been identified.

AUDIBLE SIGNAL SYSTEM: Signal system which depends on sense of hearing to attract attention.

AUTOMATIC: Self-acting. Operating by own mechanism when actuated by some impersonal influence.

BATTERY: Device that changes chemical energy into electrical energy.

BRANCH CIRCUIT: Circuit which supplies a number of outlets for lighting or appliances.

BURRS (CONDUIT): Rough edges of metal which result when cutting conduit. These must be removed to prevent damage to wire insulation.

BUS BAR: A heavy solid conductor at main power source to which branch circuits are connected.

BX: Term commonly used to identify armored cable.

CABLE, ELECTRICAL: Conductor which consists of two or more insulated wires grouped together in overall covering.

CANDLEPOWER: Term which expresses intensity of a beam of light.

CIRCUIT: Path of electric current from a source (generator) through components (such as electric lights), back to the source.

CIRCUIT BREAKER: Electromagnetic or thermal (heat operated) device that opens circuit when current in circuit exceeds predetermined amount. Circuit breakers can be reset.

CLOCKWISE: Moving in same direction as hands of a clock.

CLOSED CIRCUIT: Electrical circuit which provides path for flow of current.

CLOSING A CIRCUIT: Placing the circuit in operation.

COAXIAL CABLE: Power transmission line consisting of two conductors insulated from each other.

CODE, NATIONAL ELECTRICAL: Set of rules sponsored by National Fire Protection Association under auspices of American National Standards Institute. Purpose: To safeguard persons, buildings, contents, from hazards arising from use of electricity. Compliance will result in installations essentially free from hazard, but not necessarily efficient or adequate. The Code is not intended as a manual of instruction for untrained persons. The National Electrical Code book which is ordinarily revised every three years, is available from the National Fire Protection Association, 60 Battery-march St., Boston, Mass. 02110.

COLOR CODING: Identifying conductors by color to make sure the hot or current-carrying wires will be connected to hot wires, and the neutral wires will run in continuous circuits back to the ground terminals.

COMMON GROUND CONNECTION: Location where two or more continuous grounded wires terminate.

CONCEALED: Not readily accessible because of the structure or finish of a building.

CONDUCTOR: Material through which an electric current can flow easily. Copper wire as used in most house wiring, is a good conductor. Aluminum is also a good conductor, and is frequently used in transmission lines. Electrical energy is transferred by means of movement of free electrons that move from atom to atom inside conductor.

CONDUCTOR, BARE: Conductor which has no covering or insulation.

CONDUIT: Metal or fiber pipe or tube used to enclose electrical conductors.

CONNECTOR, SOLDERLESS: Device which establishes connection between two or more conductors by means of mechanical pressure, without using solder.

CONVENIENCE OUTLET: Electrical outlet which is part of general lighting circuit; outlet not intended for use of heavy equipment.

COUNTERCLOCKWISE: Moving in a direction opposite to the path followed by the hands of a clock.

DEVICE (ELECTRICAL): Unit or component designed to carry but not to consume current. Examples--receptacle, switch.

DIRECT CURRENT: Current which flows in only one direction. Batteries (storage, dry cell) are important sources of direct current.

DRY CELL: A nonspill type of cell which produces electricity by chemical action.

DUTY, CONTINUOUS: Requirement of service that requires operation at substantially constant load for an indefinite period of time.

ELECTRIC CURRENT: Flow of electrons through a conductor. To determine the amount of electricity (number of electrons) flowing through the conductor, it is necessary to measure the current flow. The rate of flow is determined by using the term AMPERE as the unit of measurement.
  A unit quantity of electricity is moved through an electric circuit when one ampere

of current flows for one second of time. This unit is called a COULOMB. The coulomb is to electricity, as the gallon is to measuring water.

ELECTRIC SERVICE PANEL: Main panel or cabinet through which electricity is brought into the building and distributed to branch circuits. Contains main disconnect switch for entire wiring system, also circuit breakers or fuses for individual circuits.

ELECTROLYTE: Solution of a substance which is capable of conducting electricity. This may be in form of liquid or paste.

ELECTROMAGNET: Magnet made by passing current through coil of wire wound on soft iron core.

ENERGIZE: To apply electrical voltage to or through.

EXPOSED (applied to house wiring): Designed for easy access.

FEEDER: Circuit conductor between service equipment or generator switchboard, and branch-circuit overcurrent device (fuse or circuit breaker).

FINGER-TIGHT: Tightened with the fingers.

FISH TAPE: Flat spring steel wire used to pull "fish" wires through conduit or walls.

FITTING: Accessory such as a bushing, or locknut used on wiring system intended primarily to perform a mechanical rather than electrical function.

FLEXIBLE: Something that can be bent without breaking.

FLEXIBLE CABLE OR CORD: Conductor made of a number of strands of wire of small diameter.

FOOTCANDLE: Measurement of light. Amount of illumination when one lumen (see lumen) falls on one square foot of surface.

FOOTLAMBERT: Measurement of light. Brightness of a surface which emits or reflects one lumen (see lumen) per square foot of its surface.

FUSE: Safety device inserted in series with circuit. Contains metal which will melt or break when current is increased beyond specific value for definite period of time.

GROUND: Connection between electrical circuit or equipment and the earth, or to body which serves in place of the earth.

GROUNDED: Connected to earth or to some conducting body which serves in place of the earth. In electrical wiring systems, the ground wire is always white.

GROUNDING CONDUCTOR: Wire, in circuit (green) used as safety measure. During normal use of circuit, grounding wire is not in use. In abnormal situation (for example, when live wire accidentally comes in contact with frame of washing machine) grounding wire grounds circuit to prevent injury from electrical shock.

HORSEPOWER: Unit of electrical energy equal to 746 watts of electrical power.

HOT WIRES: Wires that carry power (may be any color except white and green).

HYDROMETER: Bulb type instrument used to measure specific gravity of a liquid (ratio of weight of battery electrolyte to same volume of water).

INSULATION: Noncurrent-carrying materials used on the outside of wires, and in the construction of electrical devices.

INSULATOR: Nonconductor used to support current-carrying conductor.

ISOLATED: Object which is not readily accessible without using special means for access.

JUNCTION BOX: Box in which conductors (wires) are joined.

KILO: Prefix meaning 1,000.

KNOCKOUT: Circular metal die-cut impression in outlet and switch boxes, not completely severed, which may be removed to accommodate wiring.

LEAD-ACID BATTERY: Active materials in a lead-acid storage battery are lead peroxide (used as positive cell) and sponge lead (used as negative plate). Electrolyte is a mixture of sulfuric acid and water.

LIGHT, MEASURING: Light is measured in candlepower, lumens, footcandles and footlamberts.

Quantity	Unit
Intensity of light . . .	Candlepower
Amount of light	Lumen
Level of light	Lumen
Level of illumination . . .	Footcandle
Brightness of surface . .	Footlambert

(See also definitions of Candlepower, Lumen, Footcandle, Footlambert.)

LIGHTING OUTLET: Outlet intended for direct connection of lampholder, lighting fixture or pendant cord terminating in lampholder.

LIVE WIRE: Current carrying wire.

LOAD, CONTINUOUS: Situation where maximum current is expected to continue for three hours or more.

LUMEN: Light measurement which indicates amount of light cast upon one square foot of the inner surface of a hollow sphere of one foot radius with such a candle in its center.

MAGNET: A substance which has the property of magnetism . . . the power to attract substances such as iron, steel, nickel. Magnets may be classified as temporary or permanent depending on their ability to retain magnetic strength after the magnetizing force has been removed.

MAGNETIZE: Converting material into a magnet by rearrangement of molecules. Providing invisible force which causes material to attract steel.

MECHANICALLY SECURE: A fastening which is rigid and so made that it will not come loose unless disturbed in unnatural manner.

METER, ELECTRIC: Device which measures electricity used. Meter Construction: Small motor is connected so it turns while electricity is being used. When only a small amount of electricity is used, the motor turns slowly; when a large amount is used it turns rapidly. The motor turns gears which operate small numbered dials.

MILLIAMMETER: An ammeter that measures current in thousandths of an ampere.

MOLECULE: Smallest particle to which a substance may be reduced and still be identified by the same name. Applies to all substances . . . solids, liquids, gasses.

NATIONAL ELECTRIC CODE: See Code.

NONCONDUCTORS: Materials through which electric current does not flow. Typical examples are glass, porcelain, rubber.

OHM: Term used to indicate amount of electrical resistance in a circuit or electrical device. Resistance is often placed in a circuit to limit the amount of current which flows, or to produce heat.

OHMMETER: Instrument for directly measuring resistance in ohms.

OPEN CIRCUIT: Circuit which does not provide complete path for electric current to flow.

OUTLET: Point of wiring system at which current is obtained to supply current-consuming fixtures and equipment.

OVERCURRENT PROTECTION: Fuse or circuit breaker used to prevent excessive flow of current.

OVERLOAD: Current demand which is greater than that for which circuit or equipment was designed.

PARALLEL CONNECTION: Electrical connection which provides more than one path for flow of electricity.

POLARIZING: Identifying wires throughout the system by color to help assure that hot wires will be connected only to hot wires and that neutral wires will run in continuous uninterrupted circuits back to the ground terminals.

PLUG, ATTACHMENT: Device which, by insertion in receptacle, establishes connection between conductors of attached flexible cord and conductors permanently connected to receptacle.

POLARITY: Identification of voltage-- negative or positive.

POLARITY: Property of an electrical circuit to have positive and negative poles.

POWER: Power, whether electrical or mechanical, pertains to rate at which work is being done. Work is done when a force causes motion. The basic unit of power measurement is the watt.

QUALIFIED PERSON: Person who is thoroughly familiar with construction and operation of apparatus and hazards involved.

RACEWAY: Channel which holds electrical conductors (wires, cables, bars).

RECEPTACLE: Contact device installed at outlet for connection of a single attachment plug.

RECEPTACLE OUTLET: Outlet where one or more receptacles are installed.

RECTIFIER: Device used to change alternating current to unidirectional (one direction) current.

REMOTE CONTROL CIRCUIT: Electrical circuit which controls another circuit through a relay or equivalent device.

RESISTANCE, ELECTRICAL: Quality of an electric circuit measured in ohms that resists the flow of current. All electrical conductors offer some resistance to flow of electric current. Conductors such as copper, silver and aluminum offer but little resistance. Examples of poor conductors (insulators) are glass, wood, paper.

RHEOSTAT: Resistor which provides various amounts of resistance. Usually installed in series with load.

ROMEX: Nonmetallic sheathed cable used for indoor wiring.

ROTOR: Portion of alternating current machine which turns or rotates.

SCREW TERMINAL: Means for connecting wiring to devices, which makes use of a threaded screw.

SERIES CONNECTION: Electrical connection where there is but a single path for electricity to flow.

SERIES WOUND: Motor or generator in which armature is wired in series with field winding.

SERVICE CONDUCTORS: Conductors which extend from street main, or from transformer to service equipment of premises being supplied with electrical service.

SERVICE EQUIPMENT: Equipment located near entrance of supply conductors which provides main control and means of cutoff (fuses or circuit breaker) for current supply to building.

SHORT CIRCUIT: Improper connection between hot wires, or between hot wire and neutral wire.

SKINNING (WIRE): Removing insulation.

SOLENOID: Electromagnetic coil that contains movable plunger.

SPECIFIC GRAVITY: Ratio of the weight of a certain volume of liquid to the weight of same volume of water. Specific gravity of pure water is 1.000. Sulfuric acid used in storage batteries has a specific gravity of 1.830, thus sulfuric acid is 1.830 times as heavy as water. Specific gravity of a mixture of sulfuric acid and water may vary with the strength of the solution from 1.000 to 1.830. When a storage battery is discharged, the sulfuric acid is depleted and the electrolyte gradually changes into water. This action provides a guide in determining the state of discharge of a lead-acid battery. Specific gravity of battery electrolyte is measured with a hydrometer.

SPECIFICATIONS: Detailed descriptions which indicate manner of installation, standards of materials, workmanship, etc.

SPLICE: Connection made by connecting two or more wires.

STATOR: Fixed portion of windings of ac machine.

STRANDED WIRE OR CABLE: Quantity of small condutor wires twisted or otherwise held together to form single conductor.

SWITCH: Device used to connect and disconnect flow of current. Switches are used only in hot wires; never in ground wires.

SYMBOLS (ELECTRICAL): Arrangement of lines, letters, etc. used on house plans to show where wiring circuits, switches, outlets, etc. are to be installed.

SYSTEM (ELECTRICAL): Electrical installation which is complete and will serve purpose for which it is intended.

TACHOMETER: Instrument which indicates revolutions per minute.

THERMAL CUTOUT: Overcurrent protective device which contains heater element in addition to fusible member (melts at certain temperature) and opens circuit.

THREE-WAY SWITCH: Type of switch. Two three-way switches are required to control a light from two different locations.

TORQUE: Turning effort or twist which a shaft provides when transmitting power.

TRANSFORMER: Device composed of two or more coils, linked only by magnetic lines of force, used to transfer energy from one circuit to another.

TRANSMISSION LINES: Conductor or system of conductors used to carry electrical energy from source to load.

TUBING: Small pipe.

UNDERWRITERS LABEL: Insurance companies, manufacturers and other interested parties support testing labs called Underwriters' Laboratories. Manufacturers submit products to be tested for safety. Products which meet safety standards may use Underwriters label of approval and be placed on market.

VOLT: Unit used to measure electrical pressure (corresponds to pounds of pressure in a water system).

VOLTAGE: A measure of electrical pressure between two wires of electric circuit.

VOLTAGE DROP: Loss of electrical current caused by overloading wires, or by using excessive spans of undersize wire. Two indications of voltage drop are dimming of lights and slowing down of motors.

VOLTMETER: Instrument used to measure electrical pressure or voltage of a circuit.

WATTMETER: Instrument used to measure electrical power in watts.

WATT: Unit of measure of electric power. (Volts times amperes equals watts of electrical energy used). One watt used for 1 hour is 1 watt hour; 1000 watt hours equal 1 kilowatt hour, unit by which electricity is metered and sold.

WIRE: Electrical conductor in form of slender rod.

WIRING DIAGRAM (ELECTRICAL): Drawing in form of symbols which shows conductors, devices, connections.

Answers to Progress Checks
found at the end of the Units.

UNIT 1 (Page 10)
1. h.
2. b.
3. i.
4. d.
5. e.
6. f.
7. g.
8. c.
9. a.

UNIT 2 (Page 18)
1. noncurrent.
2. larger.
3. 12, 14.
4. black, white.
5. black, white, red.
6. True.
7. a. Nonmetallic sheathed cable.
   b. Flexible armored cable.
8. a. Lamp or fixture cords.
   b. Appliance or heater cords.
   c. Service or power cords.
9. d.
10. Underwriters' Laboratories.

UNIT 3 (Page 23)
1. conduit.
2. water pipe.
3. pressure fittings.
4. oil, water.
5. BX.
6. in-wall.
7. number, size.

UNIT 4 (Page 28)
1. True.
2. True.
3. True.
4. True.
5. a. square.
   b. octagon.
   c. rectangular.
   d. circular.
6. knockouts.
7. d.

UNIT 5 (Page 34)
1. 2, one.
2. three, two.
3. four, three.

4. mechanical parts.
5. False.
6. three-way.
7. a. Wall outlets.
   b. Fluorescent lights.
   c. Appliances.
   d. Motor-driven equipment.

UNIT 6 (Page 38)
1. c.
2. d.
3. a.
4. f.
5. e.
6. b.

UNIT 7 (Page 46)
A. Extension Rule.
B. Auger bit.
C. Wood chisel.
D. Soldering gun.
E. Multipurpose tool.
F. Conduit bender.
G. Push-pull tape rule.
H. Stubb or close quarter screwdriver.
I. Keyhole saw.
J. Fish tape.
K. Curved jaw pliers.
L. Test light.

UNIT 8 (Page 50)
1. False.
2. True.
3. False.
4. False.
5. True.
6. False.
7. True.

UNIT 9 (Page 57)
1. 120, 240.
2. two.
3. True.
4. a. Main switch
   b. Circuit breaker.
   c. Fuse.
5. 100.
6. amperage.
7. kitchen, dining area, laundry.
8. two.
9. amperage rating.

House Wiring Simplified

Answers to Progress Checks

UNIT 10 (Page 68)
1. CODES.
2. covering.
3. do not use.
4. shock.
5. a hacksaw, metal-cutting shears.
6. bushings.
7. True.
8. hacksaw, threadless.
9. False.
10. water pipe.

UNIT 11 (Page 85)
1. color.
2. False.
3. conductor.
4. 240.
5. brass, silver.
6. False.
7. metal boxes.

8. tap.
9. d.
10. overloads, short.
11. overload.
12. proper.

UNIT 12 (No Questions)

UNIT 13 (Page 124)
1. False.
2. a. armored (BX) cable.
   b. plastic cable.
   c. Nonmetallic sheathed cable.
3. hot, ground, overloaded.
4. Grounding type.
5. template.
6. five.
7. white, hot.

UNIT 14 (No Questions)

INDEX

Index